Mobility Data Management and Exploration

Nikos Pelekis • Yannis Theodoridis

Mobility Data Management and Exploration

 Springer

Nikos Pelekis
Department of Statistics
and Insurance Science
University of Piraeus
Piraeus, Greece

Yannis Theodoridis
Department of Informatics
University of Piraeus
Piraeus, Greece

ISBN 978-1-4939-4510-8   ISBN 978-1-4939-0392-4 (eBook)
DOI 10.1007/978-1-4939-0392-4
Springer New York Heidelberg Dordrecht London

to our families,
for their support and patience...

Preface

Mobility data is ubiquitous, particularly due to the automated collection of time-stamped location information from GPS-equipped devices, from everyday smartphones to dedicated software and hardware in charge of monitoring movement in land, sea, and air. Such wealth of data, referenced both in space and time, enables novel classes of applications and services of high societal and economic impact, provided that the discovery of consumable and concise knowledge out of these raw data collections is made possible. In this book, we aim at presenting a step-by-step methodology to understand, manage, and exploit mobility data: from the collecting and cleansing data stage to their storage in Moving Object Database (MOD) engines, and, then, analyzing and mining mobility data for decision-making purposes. Privacy aspects arising from handling mobility data and emerging topics, such as semantic-aware and distributed data management, towards the Big Data era are also covered. Theoretical presentation is smoothly accompanied by hands-on experience with Hermes tool, a real MOD engine developed at InfoLab, University of Piraeus.

Book Organization

The book contains 13 chapters organized in five parts. We tried to keep each chapter self-contained to provide maximum reading flexibility.

In the first part of the book, after an introduction (Chap. 1) that sets the scene and discusses at a very high level the original information (mobility data) upon which the content of the book is built, we provide the necessary background knowledge on spatial data management and exploration (Chap. 2).

The second part is about mobility data management. Preliminaries on mobility data management, including the issue of reconstructing trajectories from recorded raw locations, are discussed in Chap. 3. Modeling, querying, and indexing Moving Object Databases (MOD) is the topic of Chap. 4, while Chap. 5 presents current real-world prototype MOD implementations.

In the third part of the book, we focus on mobility data exploration. Preliminaries on mobility data exploration, including multidimensional trajectory data analysis and alternative trajectory similarity measures, are discussed in Chap. 6. Chapter 7 describes mobility data mining in order to get sound mobility patterns, which is the core of the second part, followed by a discussion about mobility data privacy aspects in Chap. 8.

The fourth part of the book studies advanced issues that have arisen only recently, namely, semantic aspects of mobility (Chap. 9) and Big Data aspects (Chap. 10).

In the fifth part, after a short epilogue (Chap. 11), two showcase appendices complete the material of the book. Using two alternative implementation of Hermes, a real MOD implemented at InfoLab and available for research purposes to the community, we provide hands-on experience on most of the topics covered in the previous chapters, from management to exploration and semantics.

The book is supported by online material: lecture slides, solutions to selected exercises, links to web resources, etc. This material can be found at the book's website: http://infolab.cs.unipi.gr/MDMEbook.

Piraeus, Greece   Nikos Pelekis
                                                                           Yannis Theodoridis

Acknowledgments

This book is the result of a 'tour' we made in several universities during the last 3 years, delivering PhD- or MSc-level intensive short courses on the topic of the book title. This tour started from Venice 3 years ago (2010), passed from Milan (2011), KAUST (2011 and 2012), Aalborg (2011), Trento (2011), Ghent (2012), Cyprus (2012), JRC Ispra (2013), and completed this round in Venice (2013). We sincerely thank our hosts in these institutions, namely, Alessandra Raffaetà, Maria-Luisa Damiani, Panos Kalnis, Torben Bach Pedersen, Themis Palpanas, Nico Van de Weghe, Demetris Zeinalipour, and Aris Tsois, for the opportunity they offered us to reach their students and junior researchers and the pleasure we had delivering those courses.

A significant part of the material presented in this book is the result of work performed in EU research projects GeoPKDD (www.geopkdd.eu), MODAP (www.modap.org), MOVE (www.move-cost.info), DATASIM (www.datasim-fp7.eu), SEEK (www.seek-project.eu), in which we have participated since 2005. We are grateful to all partners of these projects for the hours of related brainstorming.

We would also like to thank all InfoLab members, especially Despina Kopanaki, Stylianos Sideridis, Panagiotis Tampakis, and Marios Vodas, for the collaboration and significant help in the preparation of this book.

Our thanks to Springer people, namely Melissa Fearon and Jennifer Malat, for their great help and support toward the completion of this project. Their comments and suggestions were very helpful in improving the readability, organization, and overall view of the book.

As a final phrase, our wish is that this book will turn out to be useful to researchers and practitioners working on related data management topics. Even more, we hope that students will be inspired by the content found here and make their own contributions, advancing the mobility data management and exploration field!

Contents

Part II   Mobility Data Management

Part I
Setting the Scene

Τα πάντα ρει, μηδέποτε κατά τ' αυτό μένειν—Everything changes, nothing remains still.

Heraclitus

Chapter 1
Introduction

Space and *time*: the two axes according to which our lives are evolving. Every physical object has its own location (in space), a location that may change as time passes. This is how *mobility* is formed and governs our lives. Think of what 'frozen' time would mean; but this turns out to be philosophical discussion, which is for sure beyond the scope of this book... Database industry has for years been able to efficiently support time and space, though independently (the so-called, *Spatial* and *Temporal Databases*—SDB and TDB, respectively). However, it is obvious that these two axes of information find many interesting applications, if handled in conjunction. When we have in mind applications (e.g. cadastral systems) that consider spatial objects, which may change their shape or location discretely, from time to time, then we usually call them *Spatio-Temporal Databases* (STDB) whereas those that consider continuous or at least very frequent changes of objects' locations are classified under the term *Moving Object Databases* (MOD). In the latter case, the main content of the database is the so-called *mobility data*, i.e. information about the movement of objects, which includes, at least, location and time information. In this chapter, we preview the concept of mobility data and briefly discuss what can we learn from such data collections. We summarize by discussing the transition from—stationary—spatial to mobility data management and the challenges that emerge.

1.1 About Mobility Data

So, what is mobility data? It could be a variety of information:

- Data coming from our mobile phone conversations (my provider knows that I'm located somewhere during the period of my call).
- Data recorded by GPS devices during an activity (my device records that I'm at a specific location at a specific timestamp).

N. Pelekis and Y. Theodoridis, *Mobility Data Management and Exploration*,
DOI 10.1007/978-1-4939-0392-4_1, © Springer Science+Business Media New York 2014

Table 1.1 Location-aware services with respect to range (source: 3gpp.com)

Range of positioning	Examples of location-aware services
Regional (up to 200 km)	Weather reports, localized weather warnings, traffic information (pre-trip)
District (up to 20 km)	Local news, traffic reports
Up to 1 km	Targeted congestion, avoidance advice, rural and suburban emergency services, manpower planning, information services (where are we?)
100 m (67 %) 300 m (95 %)	U.S. FCC mandate (99–245) for wireless emergency calls using network based positioning methods
75–125 m	Urban SOS, localized advertising, home zone pricing, information services (where is the nearest?)
50 m (67 %) 150 m (95 %)	U.S. FCC mandate (99–245) for wireless emergency calls using handset based positioning methods
10–50 m	Asset location, route guidance, navigation

- Data exchanged between vehicles in a *vehicular ad hoc network* (VANET) environment (facilitated by a Wi-Fi network, a vehicle 'asks' its nearby vehicles about some information, e.g. the cheapest gas station in the neighborhood).
- Data collected by *radio-frequency identification* (RFID) systems; when a RFID-equipped parcel passes a RFID reader, the information that this object passed from that location at that timestamp is recorded in a database.
- Data extracted by Wi-Fi access points; the Internet access provider knows that as long as I'm served by that access point I'm located at a specific location.
- etc.

However, it is important to make clear from the very beginning that 'location' mentioned in the previous examples varies in size and resolution; it could vary from a very narrow (for instance, locations recorded by GPS devices are points with a tolerance of a few meters) to a very broad area (for instance, locations recorded by mobile phone providers could be regions of several sq.km. area). This inaccuracy in positioning is not necessarily a problem. It depends on the application whether an area of order of sq.km. is adequate positioning or the object should be identified within no more than a few meters. For example, resolution of a few meters in mandatory for effective navigation (where the road signs and turns are in front of us); on the other hand, answers to information services, such as localized weather report may be valid for a resolution of several km (Table 1.1).

Due to its popularity and its high accuracy with respect to 'location' recorded, hereafter we focus on GPS data. In the following paragraphs, we briefly introduce GPS technology and data produced by GPS devices and, then, we present examples of real GPS datasets in order to argue about the usefulness of having efficient MOD systems and algorithms for their management and exploration.

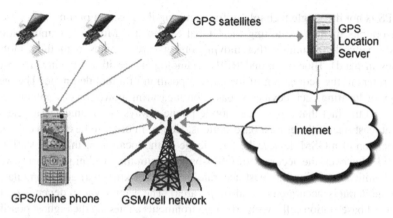

Fig. 1.1   The Assisted GPS infrastructure (source: http://www.allaboutsymbian.com)

1.1.1   Global Positioning System (GPS)

How do we collect mobility data? In particular, how to we collect GPS data? There are a number of geo-positioning technologies (most standardized in early 2000s), using either the mobile telephone network or information provided by satellites. *Global Positioning System* (GPS), including its variations *Assisted* (A-GPS) and *Differential* GPS (D-GPS), is the most popular member of the latter category.

GPS was initiated in 1978 by US Department of Defense for military purposes and it has become error-free for civilian applications since 2000. Technically, it is based on a constellation of 24 satellites monitored by 5 monitoring stations and 4 ground antennas; handled with (extremely precise) atomic clocks. The orbits of the satellites are defined in such a way that at least 5 satellites are in view from every point on the globe. A GPS receiver (onboard, on the ground) gathers information from 4 (or 3, the minimum) satellites and triangulates to position itself while at the same time fixes its (non-atomic) clock. The above setting provides to the GPS device a position accuracy of a few meters.

In particular, a GPS receiver calculates its position by measuring the time that the signals, sent by GPS satellites, take to arrive. Each satellite continually transmits messages that include: (1) the time the message was transmitted, (2) precise positioning information, and (3) the general system health and rough orbits of all GPS satellites. The receiver computes the distance to each of the satellites by using the received messages to determine the transit time of each message. These distances along with the satellites' locations are used to compute the position of the receiver. The GPS accuracy may improve down to 1 m when Assisted GPS is in use. In fact, Assisted GPS improves the triangulation calculations by getting corrections from base stations (the antennas of cellular phone network) in its vicinity, and may calculate the possible error since they also receive their "position" according to the satellites and compare it with their actual (accurate) position they are aware of. The Assisted GPS infrastructure is illustrated in Fig. 1.1.

GPS is not the single technology for acquiring the position of a moving object; other options include exploiting the Global System for Mobile Communications (GSM) network (assuming that moving objects are humans with their mobile phones in hand), Bluetooth, and RFID technologies. Position tracking using the *GSM* refers to the acquisition of the current position of a mobile phone. The technology of locating such devices is based on measuring power levels, antenna patterns, and the fact that a powered mobile phone always communicates wirelessly with at least one of the closest base stations. Hence, in order to be able to calculate the location of a GSM device, the knowledge of the location of the base station is implied. However, the accuracy of GSM-based positioning techniques varies, with cell identification being the least accurate (several sq. km. area) and triangulation being moderately accurate (several sq. m. area). The accuracy also depends on the density of base station cells, with urban environments achieving the highest possible accuracy. Related to the above discussion are the initiatives for emergency calls (911 in USA, 112 in Europe, etc.) where the call recipient tries to automatically associate a location with the origin of the call. To meet this requirement, companies have developed a variety of positioning methodologies (antenna proximity, etc.) to determine a cellular phone's location.

A quite different approach for position tracking uses Bluetooth devices and Bluetooth receivers. This technique is mostly used for indoor tracking of objects as Bluetooth receivers cover a limited area and they cannot really be used for outdoor object tracking. The movement of a Bluetooth device within an area can be tracked by considering the distances of the device from Bluetooth receivers and, as in the case of GPS, using triangulation approaches; the distance of a Bluetooth device from a specific receiver can be calculated using techniques that consider signal levels.

Similarly to the Bluetooth approach, RFID readers can locate tags within a limited area so it is hard to apply this technology for outdoor tracking of moving objects. The purpose of a RFID system is to enable data to be transmitted by a portable device, called *tag*, which is read by a RFID reader and processed according to the needs of a particular application. A typical RFID tag consists of a microchip attached to a radio antenna mounted on a substrate. A reader is needed to retrieve the data stored on a RFID tag. The data transmitted by the tag may provide identification or location information, or specifics about the product tagged, such as price, color, date of purchase, etc.

1.1.2   Format of GPS Data

Using GPS technology, a device is able to record its (time-stamped) position along with other information, such as the number of satellites that were involved in the calculations, heading, speed, etc.

Figure 1.2 lists is a snapshot of a text file in *GPS eXchange Format* (GPX), a XML-like format of data description typically used by GPS devices; a track (tagged

```
<trk>
       ...
       <trkpt lat="38.17733919" lon="23.74038222">
              <ele>862.62</ele>
              <time>2013-01-19T08:54:57.608Z</time> </trkpt>
       <trkpt lat="38.17725880" lon="23.74043843">
              <ele>1117.98</ele>
              <time>2013-01-19T08:55:21.609Z</time> </trkpt>
       <trkpt lat="38.17717291" lon="23.74039676">
              <ele>1129.98</ele>
              <time>2013-01-19T08:55:31.608Z</time> </trkpt>
       <trkpt lat="38.17707471" lon="23.74038878">
              <ele>1155.93</ele>
              <time>2013-01-19T08:55:45.584Z</time> </trkpt>
       ...
</trk>
```

Fig. 1.2 An example of GPX dataset

as <trk> in the above example) consists of a number of track points (tagged as <trkpt>), each containing the minimum required mobility information: *geographical position* and *timestamp*.

Actually, the geographical position is 3-dimensional (latitude in degrees, longitude in degrees, elevation in meters), time has an accuracy of msec, and the sampling rate (the periodicity of recordings) is one record every few seconds. It is also worth to be mentioned that the recorded position may be erroneous; especially the elevation dimension is very sensitive to errors. This should be taken into consideration when one aims to reconstruct trajectories from a dataset like this and feed a MOD for further processing and analysis. (The trajectory reconstruction problem will be discussed in detail in Chap. 3).

The <longitude, latitude> pair for representing a position on the ground of Earth is according to a polar coordinate system (*World Geodetic System*—WGS84; a universal standard) and follows the ellipsoidal representation of Earth: latitude ranges from 90° S at South pole to 90° N at North pole, while longitude spans from 180° W to 180° E, both at the antipodal meridian of Greenwich. The bad news is that several kinds of processing and analysis over mobility data assume Euclidean distance, which is not valid on ellipsoid shapes; in other words, several essential calculations should be done in meters rather than degrees. Therefore, <longitude, latitude> is usually converted to <eastings, northings> (or, more informally, <x, y>) pair, which follows a surface representation of Earth according to a 2-dimensional Cartesian coordinate system.

Converting <longitude, latitude> to <eastings, northings> is not straightforward. Actually, it is subject to the geographical zone since the objective is to transfer an ellipsoidal shape to a surface with the least deterioration of quality (position accuracy). For this purpose, there exists a universal system called *Universal Transverse Mercator* (UTM), which is very popular for non-critical applications, while for critical ones (e.g. military and cadastral systems) each country maintains its own system.

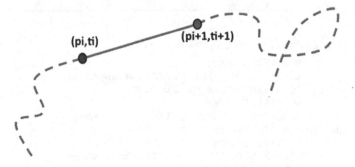

Fig. 1.3 A trajectory as a sequence of time-stamped locations of a moving object

GPS is an *absolute* location system: it uses a shared reference for all located objects. In other words, two objects placed at the same location will record exactly the same coordinates. There also exist *relative* location systems, where each object may have its own frame of reference (consider, e.g. a robot moving in a room and detecting obstacles; their location may be referred with respect to the robot's location). Hereafter, we will consider absolute location systems only.

1.1.3   Examples of Trajectory Datasets

A trajectory is a model for a motion path of a moving object (human, animal, etc.). Due to discretization, a trajectory is a sequence of sampled time-stamped locations (p_i, t_i) where p_i is a 2-dimensional point—pair (x_i, y_i)—and t_i is the recording time-stamp of p_i. In order to simulate the continuous movement of objects, a common representation of a trajectory is a 3-dimensional polyline where vertices correspond to time-stamped locations (p_i, t_i) and linear interpolation is assumed between (p_i, t_i) and (p_{i+1}, t_{i+1}). This informal definition of trajectory is illustrated in Fig. 1.3.

Although GPS devices are widespread in the real world out there, it is unusual for data owners to make their datasets publicly available since they can get added value from them (e.g. build services that exploit on this data; think of, for example, insurance companies that provide alternative contracts based on the driving behavior of their customers). Nevertheless, samples of real trajectory datasets may be found here and there and online data repositories, such as ChoroChronos.org, provide storage and processing functionality. Two indicative examples of large GPS datasets, one real and one synthetic, are presented below. These examples will be used throughout the book as running examples. Their management and exploration using a real MOD system will be also demonstrated in the final part of the book.

Example 1.1: vessels sailing on the sea. Due to International Maritime Organization (IMO) regulations, vessels with length more than a threshold are obliged to be equipped with specialized GPS devices and periodically transmit their position, along with supplementary information, in order for other vessels or port authorities to be aware of it. *Automatic Identification System* (AIS) is a tracking system used onboard by vessels and on the ground by vessel traffic services for identifying and locating vessels. AIS information along with marine radar is the primary method of collision avoidance for water transport. The content of AIS signals consists of 'static' information about the ship (identification number, name, length, draught, etc.) as well as 'dynamic' information (current location, speed, heading, destination port, etc.) while AIS receivers put a timestamp on each signal as soon as they receive it. From the above, it is clear that the critical "which/when/where" triplet of information is included in AIS. Such a real AIS dataset covers Greek Seas for a 2.5 years period and consists of order of thousands vessels, order of million trajectories and order of billion position records; admittedly, a very large dataset. A sample of this dataset, illustrated in Fig. 1.4a, is available for research purposes in ChoroChronos.org data repository.

Example 1.2: *vehicles moving on the road network of a town.* Unlike vessels, drivers in a town are not obliged to carry GPS devices on board (at least as of 2013); thus, it is not common that real mobility datasets be available for public use. On the other hand, for traffic analysis purposes, an organization may ask for volunteers to participate in the analysis by providing their mobility data. In addition, transportation scientists may use macro-simulation models to extrapolate their findings to larger populations in order to draw conclusions about traffic in a town. Alternatively, synthetic trajectory data generators may consider the road network of a real town and generate vast numbers of trajectories per request. For sake of example, let us consider a medium-sized population of 10,000 who use their GPS device in order to record their daily mobility (let us assume 2 trajectories, morning and evening, on the average) during a year. Assuming a typical sampling rate 0.2 Hz (i.e. 1 position recording every 5 s) and an average trajectory duration of 30 min, then the total number of GPS records results in order of million trajectories and order of billion position records. An illustration of a mobility dataset of this kind generated by a synthetic data generator, called *Hermoupolis*, appears in Fig. 1.4b.

1.2   What Can We Learn from Mobility Data

Having such collections of mobility data efficiently stored in a database system, we can query, analyze, and mine it in order to extract useful information or even knowledge. For instance, taking the vessels case into consideration (Example 1.1 above), we can provide analysis at individual or collective level, as such:

– *Analysis at individual level* (i.e. focusing on a single vessel): draw (a simplification of) a vessel track; calculate the similarity between its actual route and its 'expected' route; calculate its minimum distance from shore (where and when it happened); calculate the maximum number of vessels in its vicinity (e.g. within 10 n.m. radius); find whether (and how many times) it has sailed through narrow passages or biodiversity zones; find whether (and how many times) it has performed sharp changes in its direction; etc.

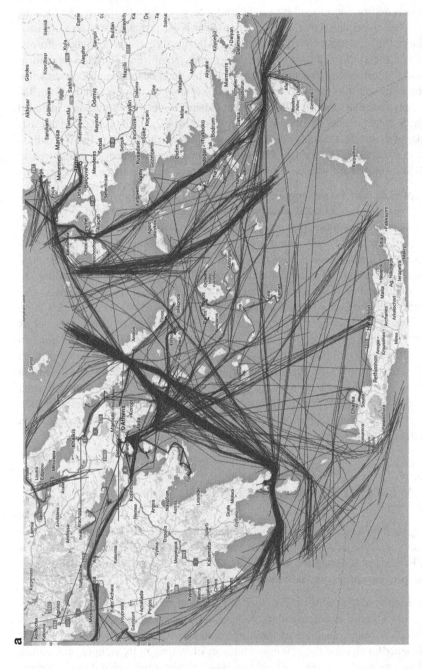

Fig. 1.4 Examples of GPS datasets: (**a**) a real AIS dataset from vessels sailing on the Greek Seas, available at ChoroChronos.org; (**b**) a screenshot of realistic trajectory data, produced by Hermoupolis, a pattern-aware synthetic trajectory data generator

b

Fig. 1.4   (continued)

- *Analysis at collective level* (i.e. involving a population of ships): find the most popular routes within the population; calculate the origin–destination matrix corresponding to this dataset, using the most popular end points of their trajectories; calculate the population's environmental fingerprint; find whether vessels with flag of convenience tend to be outliers with respect to the rest population; etc.

On the other hand, taking the vehicles case into consideration (Example 1.2 above), we can perform traffic analysis based on real-world situations (rather than simulations using traffic models) or provide personalized location-aware services. Examples include:

- *Traffic analysis*: how many cars are currently moving on the ring of the town? What is the queue per traffic light? Does the effect of "green wave" between traffic lights appear as it was originally designed? Which cars follow eco-driving recommendations? etc.

– *Location-aware services*: where is my nearest restaurant? Which are the gas stations at a maximum of 3 km distance deviation from my planned trip to my destination? I'm visiting a town; is any of my Facebook friends around? etc.

More ambitious kinds of analysis include finding mobility patterns from data (e.g. typical routes and clusters of trips), assigning a trajectory into a mobility pattern (according to a degree of similarity), or characterizing a trajectory as outlier in case it does not fit well to existing patterns, or even building flow maps or maps according to specific attributes of the database (high vs. low traffic, environmental fingerprint, etc.).

As extracted from the above example, processing of a trajectory database is essential, not only for traffic analysts to fulfill their analytical purposes but also for end users to exploit on the strength of current smartphones through *Location-based Services* (LBS) and *Location-based Social Networking* (LBSN) applications. These two families of applications (traffic analysis and LBS/LBSN, respectively) are nowadays the main driving forces for the research performed in mobility data management and exploration. Therefore, they deserve a more detailed presentation in the section that follows.

1.3   Location- and Mobility-Aware Applications

In transportation applications, a transportation network is modeled as $G = (N, E)$, where N is a set of nodes, n_i, representing e.g. road junctions, and $E \subseteq N \times N$ is a set of edges, $e_k = (n_i, n_j)$, representing e.g. routes between road junctions. This is the so-called *edge-oriented* approach for modeling a transportation network (in Chap. 3 we will discuss alternative models as well). An example of transportation network is illustrated in Fig. 1.5.

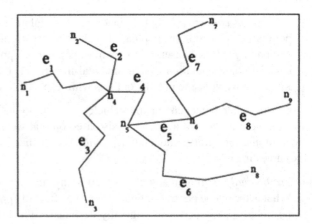

Fig. 1.5 Representing a transportation network as a set of nodes and edges between nodes

In such setting, drawing paths between an origin and a destination are important operations for traffic analysis. Actually, *routing* is a typical operation that searches for the optimal route to be followed between a source *s* and a target *t*. The cost to be minimized could be the minimum distance traveled over the network (called, "walking mode" in navigation systems), the minimum time required to travel the path (called, "driving mode") and, very recently, the most environmental friendly path (called, "eco-driving mode"). Technically, the routing problem is mapped to the well-known *shortest path* problem, extensively studied in graph theory and network analysis, with different navigation modes corresponding to different weights on the edges of the graph that maps the road network. Especially in transportation applications, a useful variation of the shortest path problem is called *time-dependent shortest path* with the difference that the cost (travel time) of edges between nodes varies with time.

Trip-planning and sequenced route queries are also variants of shortest paths, which find applications in traffic analysis. In particular, a *trip-planning* (*sequenced route*) *query* is specified by a source *s* and a target *t* as well as a sequence of facility classes, and we are searching for the shortest path from *s* to *t* visiting a set (sequence, respectively) of facilities from the facility classes. Other traffic analysis operations include calculating the *Origin–destination* (OD) matrix between points of interest, the *Congestion Index* (CI) on a road path computed as the deviation between the actual and the 'expected' (free-flow) travel time on the road, etc. Considering a population of moving objects, traffic (in general, movement behavior) analysis includes searching for *convoys*, *flocks*, and *moving clusters*.

On the other hand, LBS are services provided to the service subscribers based on their current geographic location. LBS cover a great variety of tools, from navigation and information services (routing, finding the nearest POI, "what-is-around" type location-based yellow pages) to resource management and tracing services (fleet management, administration of container goods, location-based charging, tracing of a stolen car, locating persons in an emergency situation) and advanced LBSN applications.

At individual level, plenty of mobile applications aim to record a user's movement and provide added value, from visualizing the trip to offering health services and fitness tips. Some nice examples include *Google My Tracks*, *EveryTrail*, *RunKeeper*, etc. At social networking level, some nice LBSN examples of nowadays include:

- *Facebook Places* users form a community tagging themselves in places they visit, score these places, etc. Similar for *Foursquare*, etc.
- *Google Latitude* users can see in real time where their friends are, assuming that the latter have relaxed their location privacy.
- *TripIt* is a "future LBSN" in the sense that is based on users' plans rather than their current locations.
- *Waze* is a location-based social navigation, actually a navigator based on and enriched by users' feedback.
- etc.

Of course, location-awareness also raises threats apart from opportunities. There is an increased potential for privacy violations, including the release of sensitive

Table 1.2 Taxonomy of location-aware applications with respect to the mobility of involved objects

Reference object (user) / Database objects	Stationary	Mobile
Stationary	– Routing (and constrained routing) – What is around – Find-the-nearest	– Time-dependent routing – Guide-me – Get-together
Mobile	– Find-me	

location information, user's re-identification through her (or her friends') location(s), etc. Privacy preservation is an orthogonal aspect to mobility data management and exploration and brings novel challenges in the research agenda (more on privacy in Chap. 8).

A taxonomy of location-aware applications, focusing on the objects that are involved, is provided in Table 1.2. It is built upon the mobility behavior (stationary vs. mobile) of the two "actors" involved in such applications: on the one hand, it is the reference object (the user who requests the service) and, on the other hand, it is the database objects that are searched (POIs, social network friends, etc.). For example, *find-the-nearest* service is classified as <stationary, stationary> since both the reference (the requester) and the database objects (the requested POIs) are assumed to be stationary, whereas *get-together*, e.g. in a Foursquare-like application, is classified as <mobile, mobile> since both, the user who initiates the service and her 'friends', are assumed to be moving.

At this point, it is interesting to note that the vast majority of current (as of 2013) commercial applications are classified, according to the taxonomy of Table 1.2, as either <stationary, stationary> or <stationary, mobile>. This is not strange at all; mainly, it is so due to the lack of efficient MOD management in commercial database systems.

Let us focus on the basic applications, classified as <stationary, stationary> or <stationary, mobile> or <mobile, stationary> in Table 1.2:

– *Routing* (and its time-dependent version) searches for the optimal route to be followed between a departure and a destination point. The cost to be minimized could be according to the travelling mode ("walking", "driving", etc.) as discussed earlier.
– *Constrained routing* is a variation of "routing", where additional constrains are enabled. Constrains force the object to pass through members of specified sets of candidate points, e.g. *find the best route from office to home constrained to pass from a bank ATM and a gas station*. Technically, it can be studied under the shortest path perspective (to be discussed in Chap. 4).
– *What-is-around* is also a popular service that retrieves and displays all POI's located in the surrounding area (usually rectangular or circular) with respect to a reference location (typically, the requester's current location). Technically, it is as simple as a typical range query in spatial database management (to be discussed in Chap. 2).

- *Find-the-nearest* retrieves and displays the (k-)nearest POI(s) with respect to a reference location. Technically, it is a spatial query similar (though not identical) to the typical k-NN query in spatial database management; the difference is that distance is calculated with respect to the network rather than free space (i.e. network vs. Euclidean distance).
- *Guide-me* can be thought of as the *continuous* version of *routing*: the LBS server continuously monitors the location of the requester and, if necessary, updates her routing as soon as she deviates from the previously planned path.
- *Find-me* is the inverse of *guide-me*: the reference object (the requester) remains stationary and the data objects (e.g. the 'friends' of the requester) are routed to meet her (note that the destination point—the requester's location—does not change during routing).

On the other hand, advanced (next-generation) LBS assume mobility of both reference and database objects (hence, classified as <mobile, mobile>). For instance, in *get-together*, both the reference object (the requester) and the database objects (e.g. the 'friends' of the requester) are moving. The goal of an application scenario could be that one or more of the database objects should 'meet' the reference object (or vice versa) as soon as possible. A possible methodology to address this problem is data objects to be routed to a point where the reference object is estimated to arrive after a short time interval, and this "meeting point" is periodically refreshed. Obviously, the challenge here is a successful *future location prediction*, which is one of the most challenging mobility data mining topics (to be discussed in Chap. 7).

1.4 Adding Mobility in Spatial Database Systems

So far, we have outlined a variety of location-aware applications, from basic LBS (e.g. routing) to advanced LBSN (Facebook Places, Google Latitude, etc.). In the following paragraphs we highlight the most critical operations that are essential for their efficient support from the database management perspective. They are grouped in (1) spatial database and graph- (or network-) based and (2) trajectory management operations.

On the one hand, spatial database and graph- (or network-) based operations at least include *window* and *(k-)NN search* (searching for data objects within a window or with respect to their distance from a reference object), *routing* (finding the optimal route between two points on the network, perhaps, taking a number of constraints into consideration), etc. On the other hand, trajectory management operations at least include *trajectory update* (updating the trajectory of a moving object by adding a new coming position), *trajectory map-matching* (correcting the trajectory of a moving object in order for it to be valid with respect to the underlying network), *trajectory-based search* (searching over a trajectory database—search operations include range, NN, trajectory similarity search, etc.), *trajectory projection* (estimating the location of a moving object at a future timestamp taking current and, perhaps, past information into consideration), etc.

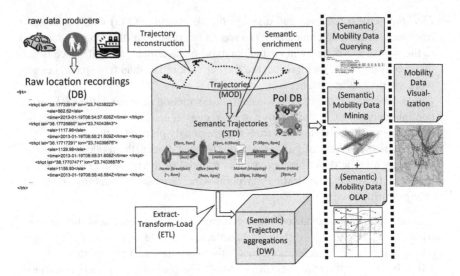

Fig. 1.6 Architecture for efficient mobility data management and exploration

In general, the big picture we exploit on this book is illustrated in Fig. 1.6. According to Fig. 1.6, the telecommunications network provides raw unprocessed data (e.g. GPS recordings), which, in turn, is processed and transformed in mobility data (e.g. trajectories of moving objects) stored in MOD systems (in their detailed form) and Data Warehouse systems (in aggregations over detailed information, respectively). The stored information can be queried as well as provide input for advanced analysis, such as multi-dimensional (OLAP) analysis and data mining, the output of which is appropriately visualized to end users.

Key questions arise from the above architecture:

– How to reconstruct trajectories from raw GPS logs? How to store trajectories in a database system?
– What kind of analysis is suitable for mobility data (in particular, trajectories of moving objects)? How does infrastructure (e.g. road network) affect this analysis?
– Which patterns/models can be extracted from them (for example, clusters, frequent patterns, anomalies/outliers, etc.)? How to compute such patterns/models efficiently?
– How to protect privacy—user anonymity? What is the tradeoff between privacy protection and quality of analysis?
– Which are the semantics hidden into the mobility data we gather? How can we make them "first-class citizens"?
– What if mobility-aware applications become so widespread, so that extremely large volumes of data bomb servers? What is the position of mobility data management and exploration in the big data era?

The above give only a few hints of what will be studied in the following chapters of this book.

1.5   Summary

Mobility data management and exploration is nowadays a hot research topic, mainly due to the wide spread of GPS devices (in smartphones, navigation systems, etc.) and related applications. In this chapter, we introduced to the concept of mobility data and briefly discussed what can we learn from mobility data collections, from typical database searching to advanced data mining. Focusing on traffic analysis and LBS/LBSN, which are the core applications for mobility data, we highlighted the technical challenges from the data management perspective. The chapters that follow will try to answer the questions that arise from this discussion.

1.6   Exercises

Ex. 1.1. Examine the dataset of Fig. 1.2 and find possible errors in the GPS measurements, if exist. Take into consideration that the GPS owner is a human walking on a mountain. By the way, in which country, close to which town, is this mountain located?

Ex. 1.2. Using one of the off-the-shelf mobile applications for tracking, record a few tracks of your own, where at least two of them correspond to the same route (e.g. home to office). Store the tracks in .gpx and .csv formats. Perform a simple processing by identifying (a) the average sampling rate and (b) the min/max extent of the main attributes: timestamp, longitude, latitude, altitude.

Ex. 1.3. Search the web for a lon/lat to x/y converter, appropriate for your geographical zone. Use this software in order to convert the positions you recorded in Ex. 1.2 from polar to Cartesian coordinate system. As you did in Ex. 1.2, identify the min/max extent of x and y attributes.

Ex. 1.4. Search in the literature for definitions of trip-planning and sequenced route queries. What are their differences? Then, focus on a home-to-office track of your own recorded in Ex. 1.2. Make a scenario where you would have to deviate from your typical home-to-office path in order to visit a couple of facilities (gas station, bank ATM, etc.).

Ex. 1.5. Search in the literature for definitions of flock and convoy queries. What are their differences? Do any of the tracks you recorded in Ex. 1.2 participate in a flock or convoy?

1.7   Bibliographical Notes and Online Resources

Koubarakis et al. (2003) was one of the earliest efforts to study STDB, from modeling to implementation aspects, as the final result of the pioneering "ChoroChronos" EU funded research project. Partially resulted by the same project, (Güting and

Schneider 2005) was the first (and still remains a "must-read") monograph in the field of MOD. A few years later, as a result of another EU funded basic research project, (Giannotti and Pedreschi 2008) discussed mobility databases, focusing on mining and privacy challenges. Very recently, (Renso et al. 2013) covered modeling, management and reasoning aspects on mobility data. A short introduction to the concept of spatio-temporal (trajectory) data is presented in (Frentzos et al. 2009).

Location systems, including GPS, Enhanced 911, etc. are surveyed in (Hightower and Borriello 2001). Especially for GIS, (Hofmann-Wellenhof et al. 2001; Djunkic and Richton 2001; Bajaj et al. 2002; Kaplan and Hegarty 2005) are very good references. Moreover, plenty of resources about GPS technology, GPX file visualization, conversion between polar and Cartesian coordinate systems, etc. can be found on the web. Examples include http://www.gpsvisualizer.com, http://cs2cs.mygeodata. eu, http://projnet.codeplex.com, etc. GPS is owned and operated by US government and is currently the single location system with global coverage; parallel initiatives are also under preparation by other states with Galileo (European Union), Glonass (Russia), and Beidou (China) being the most prominent. More details about these systems can be found in Wikipedia.org and other resources. Departing from GPS, a reference work for indoor positioning is (Xiang et al. 2004). VANETs are surveyed in (Hartenstein and Laberteaux 2008). Especially, data management issues in VANET are discussed in (Delot and Ilarri 2012).

Flow maps are used in Cartography to show the movement of objects from one location to another (number of people in a migration, etc.). A method for generating flow maps given a set of nodes, positions, and flow data between the nodes in presented in (Phan et al. 2005). Popular shortest path algorithms include Dijkstra's (Dijkstra 1959) and A* (Hart et al. 1968). Time-dependent shortest path algorithms for transportation applications are discussed in (Orda and Rom 1990; Pallottino and Scutella 1998). Routing according to alternative driving modes including eco-friendly driving is evaluated in (Ericsson et al. 2006; Guo et al. 2012; Andersen et al. 2013).

Trip-planning queries are evaluated in (Li et al 2005). Sequenced route queries in Euclidean and network-constrained space are studied in (Sharifzadeh et al. 2008) and (Chen et al. 2011), respectively. Congestion Index (CI) and other traffic parameters are evaluated in (Taylor et al. 2000). Flocks and convoys of moving objects are studied in (Gudmundsson and van Kreveld 2006) and (Jeung et al. 2008), respectively. Moving clusters are proposed in (Kalnis et al. 2005) whereas (Spiliopoulou et al. 2006) study transitions in moving clusters (e.g. appearance, disappearance, splitting, absorbing) between consecutive time points, although in a difference domain (that of document collections).

ChoroChronos.org, the online trajectory data repository responsible for the illustration in Fig. 1.4a, is outlined at (Pelekis et al. 2011) while the pattern-aware synthetic trajectory data generator, called Hermoupolis, a screenshot of which is illustrated in Fig. 1.4b, was proposed in (Pelekis et al. 2013). The LBS taxonomy presented in Table 1.2 was proposed by (Frentzos et al. 2007). More than one hundred of LBSN (active in 2013) are listed in (Schapsis 2013). Privacy aspects related to LBSN are discussed in (Ruiz Vicente et al. 2011). Trajectory update in real time is studied in (Lange et al. 2011). In that work, the authors propose a family of trajectory tracking protocols enabling several tradeoffs between computational cost, communication cost, and reduction of trajectory information.

References

Andersen O, Jensen CS, Torp K, Yang B (2013) EcoTour: reducing the environmental footprint of vehicles using eco-routes. In: Proceedings of MDM

Bajaj R, Ranaweera SL, Agrawal DP (2002) GPS: location-tracking technology. IEEE Comput 35(4):92–94

Chen H, Ku WS, Sun MT, Zimmermann R (2011) The partial sequenced route query with traveling rules in road networks. Geoinformatica 15(3):541–569

Delot T, Ilarri S (eds) (2012) Data management in vehicular networks. Transp·Res Part C Emerg Technol 23:1–124

Dijkstra EW (1959) A note on two problems in connexion with graphs. Numer Math 1:269–271

Djunkic GM, Richton RE (2001) Geolocation and assisted GPS. IEEE Comput 34(2):123–125

Ericsson E, Larsson H, Brundel-Freij K (2006) Optimizing route choice for lowest fuel consumption-potential effects of a new driver support tool. Transp Res Part C Emerg Technol 14(6):369–383

Frentzos E, Gratsias K, Theodoridis Y (2007) Towards the next generation of location-based services. In: Proceedings of W2GIS

Frentzos E, Theodoridis Y, Papadopoulos AN (2009) Spatio-temporal trajectories. In: Liu L, Özsu TM (eds) Encyclopedia of database systems. Springer, New York, pp 2742–2746

Giannotti F, Pedreschi D (eds) (2008) Mobility, data mining and privacy—geographic knowledge discovery. Springer, New York

Gudmundsson J, van Kreveld M (2006) Computing longest duration flocks in trajectory data. In: Proceedings of GIS

Guo C, Ma Y, Yang B, Jensen CS, Kaul M (2012) EcoMark: evaluating models of vehicular environmental impact. In: Proceedings of SIGSPATIAL/GIS

Güting RH, Schneider M (2005) Moving object databases. Morgan Kaufmann, San Francisco

Hart PE, Nilsson NJ, Raphael B (1968) A formal basis for the heuristic determination of minimum cost paths. IEEE Trans Syst Sci Cybern 4(2):100–107

Hartenstein H, Laberteaux KP (2008) A tutorial survey on vehicular ad hoc networks. IEEE Commun 46(6):164–171

Hightower J, Borriello G (2001) Location systems for ubiquitous computing. IEEE Comput 34(8):57–66

Hofmann-Wellenhof B, Lichtenegger H, Collins J (2001) Global positioning system: theory and practice, 5th edn. Springer, New York

Jeung H, Yiu ML, Zhou X, Jensen CS, Shen HT (2008) Discovery of convoys in trajectory databases. In: Proceedings of VLDB

Kalnis P, Mamoulis N, Bakiras S (2005) On discovering moving clusters in spatio-temporal data. In: Proceedings of SSTD

Kaplan ED, Hegarty C (2005) Understanding GPS: principles and applications, 2nd edn. Artech House, Boston

Koubarakis M, Sellis TK, Frank AU et al (2003) Spatio-temporal databases—the CHOROCHRONOS approach. Springer, New York

Lange R, Dürr F, Rothermel K (2011) Efficient real-time trajectory tracking. VLDB J 20(5):671–694

Li F, Cheng D, Hadjieleftheriou M, Kollios G, Teng SH (2005) On trip planning queries in spatial databases. In: Proceedings of SSTD

Orda A, Rom R (1990) Shortest-path and minimum-delay algorithms in networks with time-dependent edge-length. J ACM 37(3):607–625

Pallottino S, Scutella M (1998) Shortest path algorithms in transportation models: classical and innovative aspects. In: Marcotte P, Nguyen S (eds) Equilibrium and advanced transportation modelling. Kluwer, Boston, pp 245–281

Pelekis N, Stefanakis E, Kopanakis I, Zotali C, Vodas M, Theodoridis Y (2011) ChoroChronos.org: a geoportal for movement data and processes. In: Proceedings of COSIT

Pelekis N, Ntrigkogias C, Tampakis P, Sideridis S, Theodoridis Y (2013) Hermoupolis: a trajectory generator for simulating generalized mobility patterns. In: Proceedings of ECML-PKDD

Phan D, Xiao L, Yeh R, Hanrahan P, Winograd T (2005) Flow map layout. In: Proceedings of InfoVis

Renso C, Spaccapietra S, Zimányi E (eds) (2013) Mobility data: modeling, management, and understanding. Cambridge University Press, Cambridge

Ruiz Vicente C, Freni D, Bettini C, Jensen CS (2011) Location-related privacy in geo-social networks. IEEE Internet Comput 15(3):20–27

Schapsis C (2013) Location based social networks, location based social apps, and games—links. http://bdnooz.com/lbsn-location-based-social-networking-links

Sharifzadeh M, Kolahdouzan M, Shahabi C (2008) The optimal sequenced route query. VLDB J 17(4):765–787

Spiliopoulou M, Ntoutsi I, Theodoridis Y, Schult R (2006) MONIC: modeling and monitoring cluster transitions. In: Proceedings of SIGKDD

Taylor M, Woolley J, Zito R (2000) Integration of the global positioning system and geographical information systems for traffic congestion studies. Transp Res Part C Emerg Technol 8(1–6):257–285

Xiang Z, Song S, Chen J et al (2004) A wireless LAN-based indoor positioning technology. IBM J Res Dev 48(5/6):617–626

Chapter 2
Background on Spatial Data Management and Exploration

Before studying mobility data, we have to make a short tour at the (stationary) spatial domain. For decades, spatial information has been studied thoroughly; from Cartography and Geodesy to Geographical Information Systems (GIS) and Spatial Database Management Systems (SDBMS); this is justified due to its importance and ubiquity in our everyday lives. The Database community has followed the paradigm of extended DBMS and provided inherent spatial functionality in geographical data collections by developing spatial data types, operators and methods for querying, as well as indexing techniques. At the exploration level, multi-dimensional online analytical processing (OLAP) and knowledge discovery in databases (KDD) have attracted excellent results at the spatial domain. In this chapter, we review spatial database management (modeling, indexing, query processing) and exploration aspects (data warehousing and OLAP analysis, data mining), followed by a short discussion on data privacy aspects. This is essential knowledge in order for the reader to get familiar with background terms and notions during the corresponding discussion in the mobility data domain, in the chapters that will follow.

2.1 Spatial Data Modeling

From modeling perspective, in the geographical space, we can recognize a set of basic spatial entities: *points*, *lines*, and *surfaces* or *polygons* (in several variations); see the respective illustrations in Fig. 2.1. These entities are assumed to be 2-dimensional since we are interested in the (x, y)- plane. If one is interested in the 3-dimensional (x, y, z)- space then these entities are still valid, also with the addition of a fourth entity, namely *volume*.

Exploiting on these types of spatial information, let us sketch a spatial database, call it *Countries-and-Cities*, to be used as a running example for the discussion in this section.

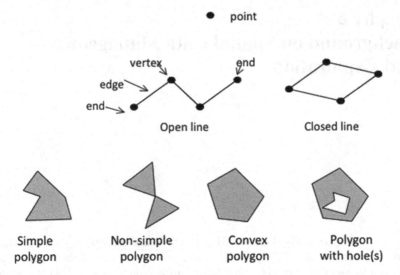

Fig. 2.1 Spatial entities in geographical space

Example 2.1. *Countries-and-Cities* **database.** *Countries-and-Cities* database is designed to support a geographical map consisting of countries (name, population, area, shape, etc.) and cities (name, population, shape, etc.). We recognize entities (a country named Greece, a city named Athens), relationships between entities (Athens is capital of Greece), alphanumeric (name of a country) vs. spatial attributes (geometry of a country, of a city), or even functions over spatial attributes of a single entity (the area of a country can be defined over its geometry) or of multiple entities (the length of borderline between two countries can be defined over their two geometries).

The question that arises is how can we model and efficiently manage information such as that of Example 2.1. Considering the different options of spatial entities presented in Fig. 2.1, 'Country' could be of polygon shape and 'City' could be of point shape. Figure 2.2 illustrates the design of *Countries-and-Cities* database at (a) conceptual and (b) logical level. The design at the conceptual level, which follows the *Extended Entity-Relationship* (EER) model, shows that we recognize two entities (Country, City) and one relationship between entities (capital-of). According to the design at the logical level, which follows the *Object-Relational* (OR) model, the database schema consists of two relations (Country, City).

Between spatial entities, we can define several relationships based on their relative positions, with *directional* and *topological* relationships being the most popular ones. Directional relationships could be absolute with respect to a Cartesian coordinate system (north-of, south-west-of, etc.) or relative with respect to a point of reference or another entity. On the other hand, topological relationships are based on the topology of entities in the geographical space and are invariant to topological transformations (shift, rotation, scaling).

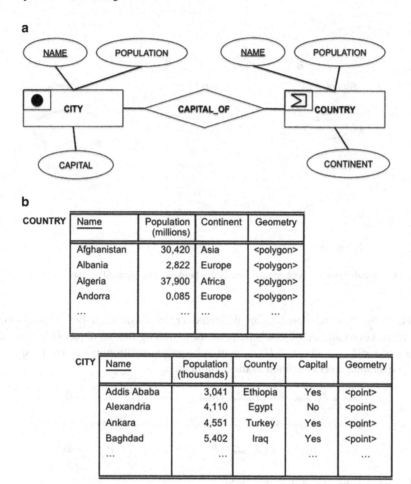

Fig. 2.2 The design of *Countries-and-Cities* database at (**a**) conceptual and (**b**) logical level

A quite popular set of topological relationships between regional objects is illustrated in Fig. 2.3 and includes:

- *Disjoint*: two regions A and B are disjoint iff A covers space which has nothing in common with the space covered by B;
- *Contain* (and its reverse, *Inside*): region A contains region B iff B is found inside A and their boundaries share no common space;
- *Cover* (and its reverse, *Covered by*): region A covers region B iff B is found inside A and their boundaries share common space;
- *Overlap*: two regions A and B overlap iff A and B share common space, A also covers some space outside B, and B also covers some space outside A;
- *Meet*: two regions A and B meet iff A and B share common space, which is found only at A's and B's boundaries;
- *Equal*: two regions A and B are equal iff A and B cover exactly the same space.

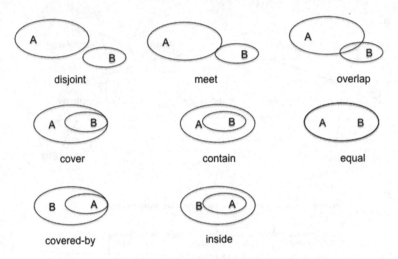

Fig. 2.3 Formalization of topological relationships between two regional objects

Formally, these relationships are defined over the possible combinations of nine set intersections among objects' components, assuming that an object O is defined as a triple $<O_{interior}, O_{boundary}, O_{exterior}>$ of sub-spaces partitioning the entire space. As an example, *meet* is defined as the combination:

$$A_{interior} \cap B_{interior} = \emptyset, A_{interior} \cap B_{boundary} = \emptyset, A_{interior} \cap B_{exterior} \neq \emptyset$$

$$A_{boundary} \cap B_{interior} = \emptyset, A_{boundary} \cap B_{boundary} \neq \emptyset, A_{boundary} \cap B_{exterior} \neq \emptyset$$

$$A_{exterior} \cap B_{interior} \neq \emptyset, A_{exterior} \cap B_{boundary} \neq \emptyset, A_{exterior} \cap B_{exterior} \neq \emptyset$$

and so on, for the rest of the relationships.

The above relationships assume regional objects, e.g. polygons. In a similar way, a (more narrow) set of relationships can be defined between linear objects (e.g. *intersect*), between one regional and one linear object (e.g. *cross*), between one regional and one point object (e.g. *inside*), and so on. In our running example, Canada and Brazil are *disjoint* whereas US and Mexico *meet*, Buenos Aires is *inside* Argentina, etc.

Queries over spatial databases obviously focus on the geometry of entities. Typical spatial queries are the following:

- *Point query* (input: spatial dataset D, point p): find objects in D that contain p.
- *Range query* (input: spatial dataset D, region r): find objects in D that overlap r.
- *Nearest-neighbor (NN) query* (input: spatial dataset D, point p, number k): find the top-k objects in D that lie nearest to p.
- *Spatial Join query* (input: spatial datasets D_1 and D_2): find pairs (o_1, o_2) of objects in datasets D_1, D_2, respectively, that satisfy a spatial condition (usually, overlap).

In our running example, pointing a location on the map we may ask to find the country it belongs to or the city that is located nearest to it; drawing a range on the map, we may ask to find the cities that are located inside it; joining countries and cities we may ask to find the cities that are located within a buffer of 10 km width along countries' borderlines; etc.

2.2 Spatial Database Management

Modern DBMS have been extended to support spatial databases like *Countries-and-Cities*. Their extensions are based on the Abstract Data Type (ADT) functionality, provided by most DBMS vendors nowadays, which is supported by appropriate operators (methods over data types) and facilitated during query processing with appropriate indexing mechanisms.

2.2.1 Abstract Data Types

In modern DBMS, spatial objects are modeled through the ADT paradigm. For instance, PostgreSQL supports a number of geometric data types—point, line segment, box, path, polygon, circle, etc.—built upon simpler data types (Fig. 2.4).

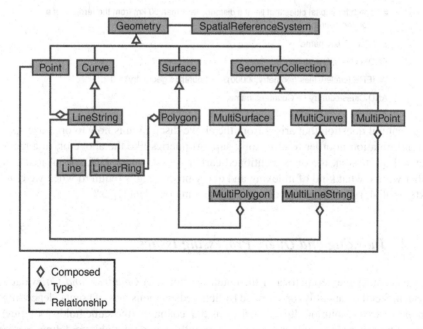

Fig. 2.4 PostgreSQL geometric data types

For instance, *line segment* is defined as a pair of points, *polygon* is defined as a sequence of points, *circle* is defined as a pair of a point and a number (radius), and so on. Moreover, the PostgreSQL ADT mechanism provides a number of methods (to find the number of points in a path, the length of a path, the center of a circle, the distance between two points, whether two boxes overlap or not, whether two line segments are parallel or not, whether a box is left of another box or not, etc.).

Having these data types in our repertoire, we may create PostgreSQL tables for *Countries-and-Cities* database as follows…

```
CREATE TABLE Countries (name varchar(40), population integer, continent
    varchar(20), geometry polygon);

CREATE TABLE Cities (name varchar(20), population integer, country varchar(40),
    capital boolean, geometry point);
```

… and search for entries with respect to alphanumeric (on the conventional attributes) or even spatial criteria (on the geometry attributes). In the following examples, we illustrate the usage of two PostgreSQL geometric operators, namely *contained in* (denoted by symbol '<@') and *distance between* (denoted by symbol '<->'):

```
# Search for cities that lie within a given rectangle

SELECT name

FROM Cities

WHERE (geometry <@ box '(476271, 420400, 479243, 420750)') = TRUE;

# Search for capital cities that lie at a distance less than 50 km from the territory of a
    foreign country

SELECT Cities.name

FROM Cities, Countries

WHERE (circle(Cities.geometry,50000) <-> Countries.geometry) = 0

AND Cities.country <> Countries.name;
```

A natural question that arises from the above discussion is how to organize spatial information in order to efficiently support queries like the above or, in a more general discussion, the ones mentioned earlier (point, range, NN, spatial join). In other words, what kind of indexing and query processing is required when we consider spatial, rather than conventional (alphanumeric) data types?

2.2.2 Indexing and Query Processing Issues

If we consider the geometries of the countries stored in *Countries-and-Cities* database, in fact they are polygons formed by hundreds or thousands of points, depending on the desired resolution. It is also obvious that countries (the same holds for cities) cannot be ordered upon their geometry. Formally, we are not able to define the '<'

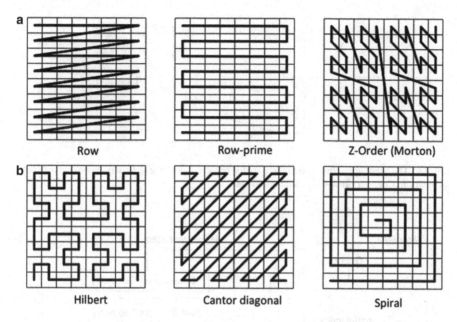

Fig. 2.5 Popular space-filling curves for mapping 2-dimensional objects in 1-dimensional space

operator so that any two entities that lie close in reality also lie close in the ordering provided by this operator, and vice-versa. If it was the case, then we could store the ordering number of each polygon and re-use the (ubiquitous) B$^+$-tree for indexing. Actually, this was one of the earliest roadmaps followed for handling spatial data and, in order to achieve it, space-filling curves, such as Hilbert ordering, were used (popular space-filling curves are illustrated in Fig. 2.5). Nevertheless, it appeared quickly that this was not the road to success and native spatial indexes had to be invented instead.

Actually, the key differences between alphanumeric and spatial data is that the latter (a) are of high complexity and (b) lack total ordering. Nevertheless, we can do something on both issues: regarding (a), we may use spatial data approximations of low complexity; regarding (b), we may adopt multi-dimensional search techniques.

Among several spatial data approximations that have been proposed, the *Minimum Bounding Rectangle* (MBR) turned out to be the most popular and widely used. As a definition, *MBR(o)* is the minimum (hyper-)rectangle, which has its sides parallel to the axes of the adopted coordinate system and entirely covers object *o*. Figure 2.6 illustrates a few examples of MBR approximations of spatial objects.

Unfortunately, though obviously, as depicted from Fig. 2.6, approximations are not identical to the original shapes they origin from. Of course, this is unavoidable in the vast majority of cases, and raises the need for a *filter-refinement* procedure to support typical spatial queries, such as the ones that were presented earlier.

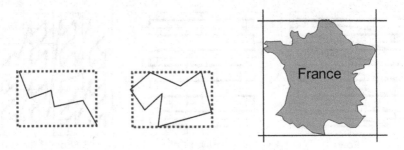

Fig. 2.6 The MBR approximation of three spatial objects: a polyline (*left*), a polygon (*center*), and a country in *Countries-and-Cities* database (*right*)

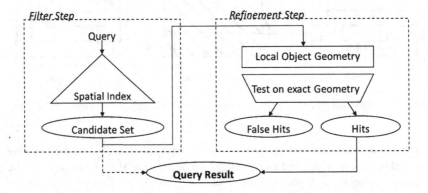

Fig. 2.7 The Filter-Refinement two-step methodology for processing spatial queries

For instance, the MBR of USA *overlaps* with the MBR of Canada but, in fact, the two countries only *meet*, with respect to the terminology of Fig. 2.3.

The Filter-Refinement methodology is a two-step process for processing a spatial query Q over a dataset D:

- *Filter step*: find a set $S' \subset D$ that contains the answer set for query Q using MBR approximations of spatial objects in D (appropriately indexed in a spatial indexing method);
- *Refinement step*: among the candidate answers, find the exact answer set $S \subset S'$ for query Q using the actual geometries of spatial objects in S'.

It is clear that, since the processing of actual geometries (usually based on computational geometry techniques) is much more expensive than the processing of the simple MBR geometry, we prefer to run the expensive part over a subset S' instead of the entire set D of spatial objects. Technically, finding the candidate answer set S' exploits on a spatial index that organizes the MBRs of spatial objects. The flow of Filter-Refinement process is illustrated schematically in Fig. 2.7.

Figure 2.7 infers that in some cases of spatial queries, it is not necessary to run the refinement step and examine the actual geometries of some of the candidate answers (see the dotted line that connects the 'Candidate Set' directly with the 'Query Result' component). In other words, an object in candidate answer set S' could directly be forwarded to answer set S without the need to examine its actual geometry. Take, for instance, the following situation: in a range query over *Countries-and-Cities* database we ask to find which countries overlap a window we draw on the map; if the MBR of a specific country fully lies inside the window then there is no need to examine the actual geometry; it is for sure that the actual geometry also lies inside the window!

According to the above filter-refinement process, query processing is facilitated by an appropriate indexing method that organizes MBR approximations of spatial objects. As already mentioned, unfortunately, we cannot adopt "total ordering" in multi-dimensional space; if this could have been done, then we could have adopted e.g. B+-trees that efficiently organize ordered information (numbers, strings, date values, etc.). Alternative solutions examined in the past to tackle this problem, researched two main directions:

1. adopt a partial ordering mechanism to transform spatial objects into numbers or ranges of numbers and then index them using B+-trees, or
2. invent novel spatial indexing techniques, mostly based on the B+-tree concept.

As already mentioned, techniques falling under the first category exploited on popular space traversal methods, such as Peano and Hilbert (see Fig. 2.5). However, it was the second category that finally was the winner in this debate, with R-tree being its most popular representative. It is not exaggeration to argue that, nowadays, R-tree is "ubiquitous" in extensible DBMS as B+-tree was (and still is) for conventional DBMS applications.

Actually, R-tree extends B+-tree ideas to multi-dimensional space. Like B+-tree, R-tree is a height-balanced tree with the index records in its leaf nodes containing pointers to the actual data objects and guarantees that the space utilization is at least 50 %. Leaf node entries are in the form (*id, MBR*), where *id* is an identifier that points to the actual object and *MBR* is the MBR approximation of the actual object. Non-leaf node entries are of the form (*ptr, MBR*), where *ptr* is a pointer to a child node, and *MBR* is the minimum bounding rectangle covering all child nodes. In a disk-based implementation, a node in the tree corresponds to a disk page and contains between m and M entries. The maximum number M of entries in each node (otherwise, *fanout*) is predefined according to the size of the node (disk page) and the size of each entry. An example of the R-tree indexing method in 2-dimensional space appears in Fig. 2.8:

In the example of Fig. 2.8, 10 data MBRs (r_1, ..., r_{10}) are organized in 4 leaf nodes (assuming fanout $M=3$), which are, in turn, organized in 2 nodes containing the respective MBRs (R_3, ..., R_6), which are, in turn, organized in a root node containing the respective MBRs (R_1, R_2), thus forming a tree of height 3.

Propagating down the tree in an appropriate way, we can answer point, range, NN, and spatial join queries very efficiently: in logarithmic cost instead of the linear

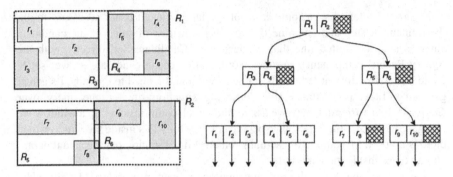

Fig. 2.8 An example of R-tree indexing technique

cost that would be the case if no index existed. For sake of completeness, we sketch the typical R-tree-based algorithms for processing range (point included as a special case), NN, and spatial join queries in spatial databases. This is because these algorithms are used as starting points for plenty of algorithms and techniques on mobility data that will be discussed in the chapters that follow.

R-tree nodes contain entries consisting of a pair (*MBR, ptr*) with the MBR approximation of the enclosed geometry and a pointer to the child node (or to the actual geometry, if the entry is at the leaf level), respectively. In range and spatial join query algorithms sketched below, starting from the root node we propagate down the R-tree(s) considering whether the node entries overlap with the geometry of interest (the query window or an entry in the second R-tree, respectively). As such, the cost is order of height of the R-tree, hence logarithmic with the number of entries. On the other hand, unlike B^+-tree, it is not unusual the case where more than one paths should be propagated in order to answer spatial queries. For example, if we perform a point query on the R-tree of Fig. 2.8 with the query point p be located close to the upper-right corner of r_8, then we need to visit both nodes under R_5 and R_6, although visiting the latter will turn out to be useless.

Algorithm R-tree-based-RangeSearch

Input: R-tree node N, query window Q

Output: answer set S

```
1.   IF N is a non-leaf node
2.       FOR each entry e in N
3.           IF e.MBR overlaps Q
4.               CALL R-tree-based-RangeSearch (e.ptr, Q)
5.   ELSE // N is a leaf node
6.       FOR each entry e in N
7.           IF e.MBR overlaps Q
8.               add e to S
```

Algorithm R-tree-based-SpatialJoin

Input: R-tree node N1, R-tree node N2 // assumption: R-trees of equal height

Output: answer set S

```
1.   IF N1, N2 are non-leaf nodes
2.       FOR each pair of entries (e1, e2) in N1xN2
3.           IF e1.MBR overlaps e2.MBR
4.               CALL R-tree-based-SpatialJoin (e1.ptr, e2.ptr)
5.   ELSE // N1, N2 are leaf nodes
6.       FOR each pair of entries (e1, e2) in N1xN2
7.           IF e1.MBR overlaps e2.MBR
8.               add (e1, e2) to S
```

In a similar fashion, NN queries are addressed as follows (p being the query point): starting from the root node, all entries are sorted according to the lower bound of their potential distance from p (called, MINDIST), and the entry with the smallest distance is visited first. The process is repeated recursively until we reach the leaf level; there, a candidate answer (NN) is found. During backtracking and in order to improve the answer, we visit only those entries whose MINDIST is smaller than the distance of the already found NN from p.

Algorithm R-tree-based-NNSearch

Input: R-tree node N, query point P

Output: object S

```
1.   nearest = INFINITY
2.   IF N is a non-leaf node
3.       FOR each entry e in N
4.           Add e in BranchList in ascending order according to MINDIST
5.       Prune BranchList with respect to MINDIST
6.       FOR each entry e in BranchList
7.           Prune BranchList with respect to MINDIST
8.           CALL R-tree-based-Range-Search (e.ptr, P)
9.   ELSE // N is a leaf node
10.      FOR each entry e in N
11.          IF distance (e.MBR, P) < nearest
12.              nearest = distance (e.MBR, P)
13.              S = e
```

For example, if we perform a NN query on the R-tree of Fig. 2.8 with the query point p be located close to the lower-right corner of R_3, we retrieve R_1, then R_3, then r_2, but it is not enough; we then have to retrieve (at least) R_6 (since its MINDIST is smaller than the distance between p and r_2), and, surprisingly, the correct answer is found there (it is r_9)! Also, please take into consideration that if we are searching for the actual geometry (instead of the MBR) that lies nearest to the query point, the algorithm should be aware of it and perhaps provide more than one candidate answers to be passed to the refinement step.

2.3   Spatial Data Warehousing

Exploring spatial data includes spatial data warehousing for OLAP analysis purposes and spatial data mining for extracting hidden knowledge. Data Warehousing (DW) has been widely investigated for conventional, non-spatial data. According to a popular definition, *a Data Warehouse is a subject-oriented, integrated, time-variable, non-volatile information system aiming at decision-making*. To rephrase the above definition, DW plays the role of a database that integrates historical information from other sources (operational databases, etc.) according to a subject, and upon the information stored in this database, multi-dimensional analysis is performed aiming at decision-making (the term OLAP, for *On Line Analytical Processing*, is widely used for this kind of analysis). As such, it is rather static database and, only periodically, it is refreshed by looking into updated sources through an ETL (for *Extract—Transform—Load*) process.

A typical DW architecture is illustrated in Fig. 2.9a: data sources (mainly, operational databases) feed the DW using an ETL (and, periodically, *Refresh*) process; the DW is also accompanied by metadata information; using an OLAP engine, different (subject-oriented) data cubes are built upon the data stored in the DW; operations upon the data cubes include multi-dimensional OLAP analysis, cross-tab querying and reporting, and, more advanced, data mining processing.

Technically speaking, a data cube is a view over the database stored in DW. Its typical structure (in a relational model terminology) consists of a centrally located *fact table* with measures and foreign keys to *dimension tables*, the so-called *star schema*. Figure 2.9b illustrates such an example for a conventional database (e.g. retail): the data cube is three-dimensional, where Time, Product and Location are the three dimensions, whereas quantity and turnover are the two measures to be analyzed. Other, more complex data cube designs include the so-called *snowflake* and *constellation* schemas; in the former, each dimension may be represented by joins over more than one tables, while in the latter, more than one fact tables are present, sharing common dimension tables.

Having such information in their hands, business analysts may perform multi-dimensional analysis over the measures and dimensions defined in the data cube. The most popular OLAP operations include *roll-up* (moving from a finer to a coarser analysis), *drill-down* (the opposite of roll-up), *slice* (performing a filtering according to a condition), and *cross-over* (looking back to the detailed information that supports the OLAP result). An example series of analyses/operations on the data cube that appears in Fig. 2.9b is the following:

– *Find the total turnover per month and per city*, a roll-up operation;
– *Especially in March, find the turnover per city*, a slice operation;
– *Especially in March, find the turnover on weekdays* vs. *weekends*, a drill-down operation.
– *Retrieve the database records that support the above result*, a cross-over operation.

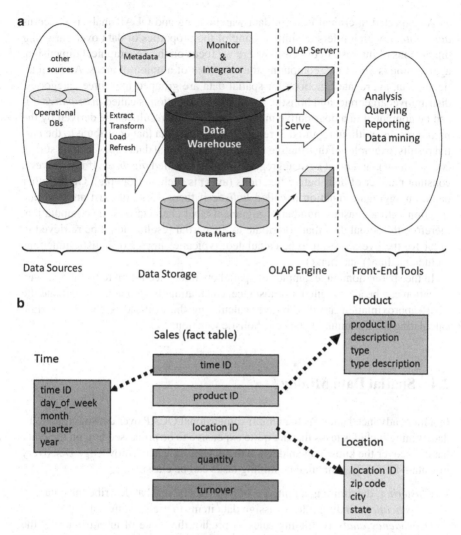

Fig. 2.9 Conventional data warehousing: (**a**) typical DW architecture; (**b**) typical data cube following star schema

Adding spatial dimensions and measures in the above paradigm results is the so-called *Spatial DW*. In correspondence to the traditional paradigm, Spatial DW include:

- At the data cube design level: spatial measures (in the fact table), spatial dimensions and spatial hierarchies (in the dimension tables);
- At the OLAP analysis level: spatial OLAP operations, i.e. operations like roll-up and slice extended to spatial predicates;
- At the data cube implementation level: appropriate indexes and efficient query processing techniques.

As expected, a critical issue in data warehousing and OLAP analysis concerns *aggregation*, which refers to summarizing of the properties of data over particular dimensions of interest. In our case, where we focus on the geographical dimension, aggregation is about e.g. computing the total area of a union of areas. As for relational data, aggregate functions for spatial data are grouped into three categories: distributive, algebraic, and holistic. An aggregate function is called *distributive* if it can be computed in a distributive manner, i.e. the final result can be derived by the partial results, without required to retrieve the actual data that contribute to the partial results; examples of distributive functions in relational data cubes include sum(), count(), min(), and max(). An aggregate function is called *algebraic* if it involves a constant number of distributive functions (avg() is such an example). On the other hand, an aggregate function is called *holistic* if there does not exist an algebraic function with a constant number of arguments that characterizes the computation, therefore the actual data that contribute to the partial results should be retrieved in order for the overall result to be calculated; typical examples of holistic functions include median() and rank().

In the spatial domain, examples of spatial operators according to the above classification can be also defined. For instance, calculating the geometric union and the MBR approximation are distributive, calculating the centroid is algebraic, while calculating the minimum distance is holistic function.

2.4 Spatial Data Mining

In a more advanced analysis than multi-dimensional OLAP over data aggregations, data mining includes tasks that are quite expensive to be processed but, on the other hand, uncover the knowledge hidden in the database (hence, Knowledge Discovery in Data—KDD). In general, data mining tasks can be classified as:

– *Clustering*: determining a finite set of implicit classes that describe the data.
– *Classification*: finding rules to assign data items to pre-existing classes.
– *Dependency analysis*: finding rules to predict the value of an attribute on the basis of the values of other attributes.
– *Deviation and outlier analysis*: searching for data items that exhibit unexpected deviations or differences from some norm.
– *Trend detection*: finding lines and curves to data to summarize the database.
– *Generalization and characterization*: obtaining a compact description of the database, for example, as a relatively small set of logical statements that condense the information in the database.

All the above types are also applicable in *spatial data mining*. In the remainder of the section we will focus on cluster analysis (also with a note on sampling, which is a very related problem) and dependency analysis in the form of co-location pattern mining.

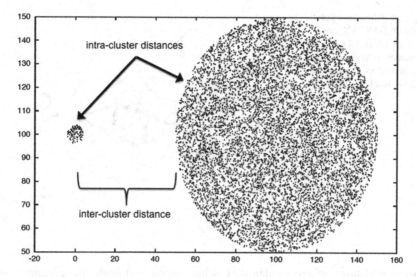

Fig. 2.10 The concepts of intra-cluster and inter-cluster distances

2.4.1 Cluster Analysis

In cluster analysis, the objective is to map a dataset of N spatial objects, $D = \{o_1, \ldots, o_N\}$, usually points, into a clustering consisting of K groups (clusters), $C = \{C_1, \ldots, C_K\}$, so as similar objects are grouped together in the same cluster and dissimilar objects are grouped in different clusters; similarity between a pair of objects is defined according to a similarity function $Sim(o_i, o_j)$. In other words, the ultimate goal is that *intra-cluster* distances are minimized and *inter-cluster* distances are maximized, whatever is the shape and density of the resulting clusters (Fig. 2.10).

To address this problem, there exist dozens of algorithms; others work in a *hierarchical* fashion, finding all possible groupings from 1 to N clusters while others result in a specific clustering of K clusters; others assume the number K of clusters to be input while others not; others are able to find and isolate outliers while others not. There is also an important classification of clustering algorithms as *partitioning* (where the target is to partition the space in K partitions and each partition corresponds to a cluster), and *density-based* (where clusters are formed according to the spatial density of objects and may take irregular shapes).

As a typical example of conventional data mining technique, K-means, perhaps the most popular clustering algorithm, is a partitioning algorithm aiming to separate the space in K distinct partitions (thus, favoring spherical clusters). It does not distinguish between clustered points and noise, so noise removal should be performed in an earlier data preparation phase. Although intuitive and generic, partitioning algorithms like K-means have some shortcomings. In detail, they are not capable of detecting clusters of arbitrary shape and they are weak in detecting meaningful

Fig. 2.11 Clustering point
data with DBSCAN: core
(in *red*; dark grey in b/w)
vs. border (in *yellow*; light
grey in b/w) vs. noise points
(in *blue*; black in b/w)

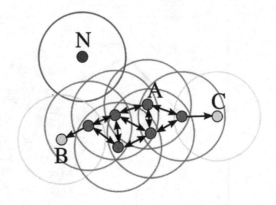

clusters in datasets where the density has big variations. Finally, these techniques
have the intrinsic limitation that the user should tune the *K* parameter, which corresponds to the number of clusters, as well as the fact that they are sensitive to noise
(actually they cannot handle it at all).

To overcome these shortcomings, the concept of density-based clustering
emerged to propose nice solutions for them. The core idea of density-based clustering algorithms is that clusters are formed in areas of higher density than the rest of
the dataset, while points in sparse areas (called noise or border points) separate the
clusters among them. This category of algorithms is robust to noise and outliers,
while they can discover clusters of various shapes, making them suitable for many
real world applications, as those working with mobility data.

The DBSCAN algorithm is one of the most well-known density-based clustering
algorithms in the literature. According to DBSCAN, a cluster is a maximal subset
of points in the database that are *density-connected*. In order to clarify the notion of
density-connected points, we need first to define the concept of *density-reachability*.
Given a distance threshold ε and an integer *minPts*, a point p_{i+1} is *directly density-
reachable* from a point p_i, if p_{i+1} belongs to the ε-*radius neighborhood* of p_i, inside
which there are more than *minPts* points (i.e. as such it is considered as a dense
(core) point). A point p_n is said to be *density-reachable* from a point p_1 w.r.t. ε and
minPts, if there exists a chain of points $p_1, \ldots, p_n$ such that p_{i+1} is directly density-
reachable from p_i, $1 \leq i < n$. The density-reachable relation is not symmetric (i.e. p_1
might not be *density-reachable* from p_n, even if the reverse is valid) as p_n might lie
on the edge of a cluster, having in its ε-radius neighborhood less than *minPts* points.
This asymmetry is repaired by the concept of *density-connectivity*. More specifically, two points p_1 and p_n are *density-connected*, if there is a point p such that both
p_1 and p_n are *density-reachable* from p. Now, the density-connected relation is symmetric. In the example presented in Fig. 2.11, points at A are core points, points B
and C are density-reachable from A (though not core, hence border points), whereas
point N is neither core nor density-reachable from a core, hence noise.

DBSCAN starts with an arbitrary point that has not been visited. If the *ε-radius neighborhood* of the selected point contains at least *minPts* points, a cluster is initiated, otherwise the point is labeled as noise. If a point is a dense part of a cluster, its *ε-radius neighborhood* is also merged to that cluster. This process continues until no further density-connected points can be found. In such a case, a new unvisited point is selected, which either initiates another cluster or it is considered noise.

Algorithm DBSCAN

Input: dataset D, neighborhood distance eps, population threshold MinPts

Output: clusters C and noise N

```
1.   C = 0
2.   FOR each unvisited point P in dataset D
3.        mark P as visited
4.        NeighborPts = regionQuery(P, eps)
5.        IF sizeof(NeighborPts) < MinPts
6.             add P to noise N
7.        ELSE
8.             C = next cluster
9.             expandCluster(P, NeighborPts, C, eps, MinPts)
```

expandCluster(P, NeighborPts, C, eps, MinPts)

```
1.   add P to cluster C
2.   FOR each point P' in NeighborPts
3.        IF P' is not visited
4.             mark P' as visited
5.             NeighborPts' = regionQuery(P', eps)
6.             IF sizeof(NeighborPts') >= MinPts
7.                  NeighborPts = NeighborPts joined with NeighborPts'
8.        IF P' is not yet member of any cluster
9.             add P' to cluster C
```

regionQuery(P, eps)

```
10.  RETURN all points within P's eps-neighborhood (including P)
```

A weak point of DBSCAN is that it is unable to discover clusters of varying density. Its successor, OPTICS solved this shortcoming, by also making one of the two parameters, i.e. the ε distance threshold, not necessary, as the user may easily set it by a maximum value implying the *maximum search radius*. According to OPTICS, objects are linearly ordered based on their spatial closeness. Additionally, for each point, a distance measure (called *reachability distance*) that represents the required density of this point to be accepted to belong to the same cluster with its neighbor is stored. The *reachability distance* of point p_1 with respect to point p_2 is actually the smallest distance, such that p_1 is directly density-reachable from p_2, provided that p_2 is a core point, otherwise it is undefined. The final outcome of the OPTICS algorithm is a so-called *reachability plot* in which the ordering of the points forms the

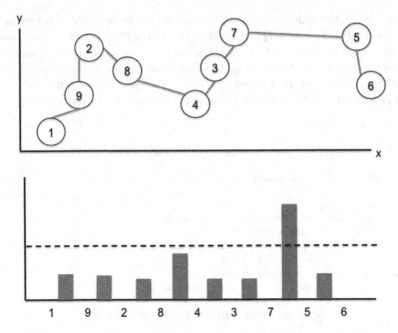

Fig. 2.12 Clustering point data with OPTICS: a set of spatial objects traversed by their pair-wise proximity (*top*) and the resulting reachability plot (*bottom*)

x-axis and the y-axis is the reachability distance of the corresponding point. This plot may be considered as a special kind of *dendrogram*, which enables the derivation of the formed clusters. Intuitively, objects within the same cluster have small values of *reachability distance*, therefore correspond to valleys in the reachability plot. Obviously, the deeper a valley is, the denser the corresponding cluster.

Figure 2.12 illustrates an example of OPTICS clustering. Starting from an arbitrary object, each one finds its nearest-neighbor among the non-visited ones (top of Fig. 2.12) and this traversal results in a reachability plot (bottom of Fig. 2.12). The dashed bar on the plot decides the grouping of objects in clusters, according the valleys produced (two in our example).

Very related to clustering is the *sampling* problem, where the goal is to select a representative subset from a large population so as the 'properties' (distribution in space, classes, patterns, etc.) of the original set are maintained in the sample. Effective sampling in spatial data should definitely take the density information into consideration. Consider, for instance, the two clusters illustrated in Fig. 2.10 above. The rightmost consists of a much higher number of points than the leftmost. On the other hand, the density (i.e. the number of points in each cluster with respect to its area) of the small cluster is larger than that of the large one. Uniform sampling cannot maintain this property, and the large cluster will be over-represented in the sample. Moreover, by uniform sampling a very small cluster is in dangerous not to be represented at all. To prevent these problems, density-biased algorithms favor dense clusters in order for them to participate with an adequate number of points in the sample.

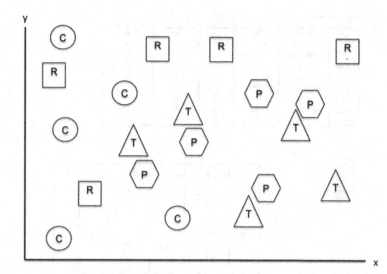

Fig. 2.13 Example of co-location pattern mining: co-location rules "T → P" and "P → T" appear frequently

2.4.2 Co-Location Pattern Mining

In spatial co-location pattern mining, the objective is to find co-location rules in a spatial database. For example, in an animal ecology database that stores locations of wild animals in the jungle, a co-location rule "lion → tiger" indicates that wherever a lion appears, a tiger also appears in lion's neighborhood with high probability. Formally, given a spatial dataset consisting of tuples <*id*, *ft*, *loc*>, where *id* is the tuple identifier, *ft* is the feature type, and *loc* is the location, and a neighbor relation *R* over locations, the goal is to find co-location rules "X → Y" that are found in the dataset with high (above a threshold) probability.

Figure 2.13 illustrates an example of co-location pattern mining. Assume a population of circles (C), rectangles (R), triangles (T), and pentagons (P): it is clear that triangles and pentagons are co-located with high probability, which results in co-location rules "T → P" and "P → T".

2.5 Data Privacy Aspects

The respect and protection of human dignity, secrecy and free growth of personality, constitute fundamental pursuits of every democratic society. Among human rights, *privacy* is perhaps the most difficult to define. Privacy can be interpreted as *the personal right of individual to choose freely how she will manage her personal information.*

SSN	name	ZIP	age	gender	Disease
123	Smith	47677	65	M	A
234	Jones	47602	63	M	A
345	Williams	47678	67	M	B
456	Clinton	47905	43	F	B
567	Jameson	47909	52	F	C
678	Newman	47906	47	F	C
789	Carlson	47605	30	M	D
890	Patterson	47673	36	M	D
901	Turner	47607	32	M	B

id	ZIP	age	gender	Disease
1	47677	65	M	A
2	47602	63	M	A
3	47678	67	M	B
4	47905	43	F	B
5	47909	52	F	C
6	47906	47	F	C
7	47605	30	M	D
8	47673	36	M	D
9	47607	32	M	B

Fig. 2.14 A hospital's dataset: original in-house (*top*) vs. sanitized published version (*bottom*)

With the rapid growth of technology, privacy aspects came in the light. Nowadays, our everyday actions leave traces (shopping transactions with credit and loyalty cards, electronic administrative transactions, health records, etc.). Several public and private bodies are by law obliged to publish available datasets for analysis purposes. A naïve approach of maintaining individuals' privacy suggests that by only removing the obvious identifiers such as identity, privacy is preserved.

However, linking data from various sources might lead to draw inferences, which are not possible to conclude from a single source. Consequently, when data holders remove identifiers only, they cannot actually guarantee the anonymity of entities whose data undergoing public release. The set of attributes that each one is not specific enough to identify individuals but through their combination re-identification of an individual can be accomplished, are called *quasi-identifiers*. Quasi-identifiers contain data, such as ZIP code, birthdate, gender, etc. Whereas, attributes that contain information that individuals want to keep secret, are called *sensitive attributes*.

Consider the following example, in which a hospital collects medical and demographic data about its patients. Should for research purposes this dataset be analyzed, the hospital is obliged to provide a "sanitized version" of it (by removing identifiers, such as name and SSN). In this way, no one is (or should be) able to identify that a patient suffers from a specific kind of illness. Figure 2.14 illustrates the dataset that the hospital finally releases. In this table, {ZIP, age, gender} are the quasi-identifiers while Disease is the sensitive attribute.

The "privacy breach" is clear in the above example. Assume that an adversary knows that Mr. Smith, aged 65 and living in ZIP 47677, is included in the dataset, then she may infer with 100 % confidence that Mr. Smith suffers from disease "A" (since there is only one record with these quasi-identifier values).

In order to protect individuals' privacy from attacks of this kind, the K-*anonymity* principle requires that, in the released dataset, the individuals should be indistinguishable with respect to the set of quasi-identifiers. Formally, a dataset is considered to be K-*anonymous* when for any given set of quasi-identifiers, a record is indistinguishable from K-1 other records. The set of records satisfying K-anonymity requirement, is called *equivalence class*. K-anonymity principle guarantees that an adversary cannot associate a particular record with a specific individual with probability greater than $1/K$, even she: (a) has gained access to the anonymized dataset, (b) is informed that a given individual is included in the dataset, and (c) is aware of the corresponding set of quasi-identifiers.

K-anonymization may be achieved through generalization, suppression or perturbation of quasi-identifier values. More specifically, *generalization* replaces original with higher-level values, taking conceptual hierarchies into consideration (where lower levels contain more information than higher levels). Note that as soon as a value is decided to be generalized, all values of this attribute are generalized, accordingly. Alternatively, *suppression* is performed by hiding the values of the attributes that do not appear in the dataset at least K times. Finally, *perturbation* replaces the actual value with a random value from the standard distribution of values for that attribute. The overall distribution of values is not affected but the individual data values are not correct. The objective of all these techniques is to provide a K-anonymized version of the dataset with the minimum deviation from the original.

Figure 2.15 is the K-anonymized version ($K=3$, 4) of the dataset illustrated in Fig. 2.14, where the generalization method has been adopted. For instance, gender values are generalized to 'undefined' while ZIP and age values are generalized to ranges. Note that in the K-anonymous version, every tuple belongs to an equivalence class together with at least K-1 other tuples. As such, continuing the previous example, the adversary now cannot infer Mr. Smith's disease, since at least K tuples exist with the same quasi-identifier values as of Mr. Smith. Moreover, it is obvious that the larger the K, the higher the deviation from the original information.

Although intuitive, K-anonymity may not be able to protect sensitive values when their diversity is low. More particular, it suffers from two types of attacks: *homogeneity* and *background knowledge* attack. The former corresponds to cases where a sensitive value occurs repeatedly in a set of tuples with the same quasi-identifier values after the anonymization of the dataset; the latter involves extra information available to an attacker, which may allow to an attacker to increase probability of being able to determine sensitive information about an individual. In the example illustrated in Fig. 2.14 (top), if the disease of id=3 was "A" instead of "B" then an adversary would infer that Mr. Smith's disease is "A" (homogeneity attack). On the other hand, if an adversary knows that Mr. Smith has very low

id	ZIP	age	gender	Disease
1	476**	60-79	M	A
2	476**	60-79	M	A
3	476**	60-79	M	B
4	479**	40-59	F	B
5	479**	40-59	F	C
6	479**	40-59	F	C
7	476**	20-39	M	D
8	476**	20-39	M	D
9	476**	20-39	M	B

($K = 3$)

($K = 4$)

id	ZIP	age	gender	Disease
1	47*	50-99	*	A
2	47*	50-99	*	A
3	47*	50-99	*	B
5	47*	50-99	*	C
4	47*	00-49	*	B
6	47*	00-49	*	C
7	47*	00-49	*	D
8	47*	00-49	*	D
9	47*	00-49	*	B

Fig. 2.15 The K-anonymous version of the hospital's dataset: $K=3$ (*top*); $K=4$ (*bottom*)

probability of having disease "B" then she could infer that most probably Mr. Smith's disease is "A" (background knowledge attack).

Finally, in the area of *Privacy Preserving Data Mining* (PPDM), there have appeared methods for modification of data before publishing, in order for sensitive information to be protected. Such techniques include *perturbation* (altering an attribute value by a new value or adding noise), *blocking* (replacing an existing value by e.g. a "?"), *aggregation* or *merging* (replacing a value by the respective of a coarser category with respect to a hierarchy), *swapping* (interchanging values of individual records), and *sampling* (releasing data for only a sample of a population).

2.6 Summary

Spatial information has been studied for years in SDBMS. In this chapter, we provided a brief overview of the topic, from modeling to management and exploration, which by nature is huge and cannot fit in a book chapter. Nevertheless, the presented material is the necessary background knowledge for the discussion that will follow in the next chapters regarding the mobility data domain.

2.7 Exercises

Ex. 2.1. Develop your own Countries-and-Cities database sketched in Example 2.1. To populate it with real information, you may look at various online resources (Wikipedia.org, Wikimapia.org, OpenStreetMap.org, etc.)

Ex. 2.2. From Fig. 2.6, it is clear that MBR approximations of spatial objects are not identical to the original shapes they origin from. however, this in not always true; there do exist objects with the property that their MBR is equal to their actual geometry. Find at least three 2-dimensional shapes with this property.

Ex. 2.3. According to the discussion that followed Fig. 2.7, there exist spatial queries that make the refinement step unnecessary for some of the candidate answers. Could you think of cases of spatial queries that make the refinement step completely unnecessary, i.e. the candidate answer set S' is by definition equal to the actual answer set S?

Ex. 2.4. Using the example R-tree of Fig. 2.8, perform (1) a point query where the reference point p is located in the center of R_1, (2) a range query where the query window r is equal to the intersection $R_5 \cap R_6$, and (3) a 1-NN query where the reference point p is located at the lower-right corner of R_2. What are the lessons learned?

Ex. 2.5. Consider the example data cube illustrated in Fig. 2.9b: (a) extend it to spatial data cube by adding a 'geometry' attribute in the spatial dimension; (b) over the 'geometry' attribute, define the aggregate functions "minimum bounding box", "region area", "region perimeter", "center of gravity" and argue for each of them whether it is distributive, algebraic, or holistic.

Ex. 2.6. According to the discussion about OPTICS, the reachability plot can be used to produce a dendrogram that provides clustering alternative, from a single cluster including all entities to a cluster per entity. Display the dendrogram that corresponds to the reachability plot illustrated in Fig. 2.12 (bottom).

Ex. 2.7. In the dataset of cities you populated for the purposes of Ex. 2.1, find groups and outliers using DBSCAN and OPTICS. What are the lessons learned? For instance, how did you manage to set appropriate values in the parameters of the algorithms?

Ex. 2.8. From the Countries-and-Cities database developed in Ex. 2.1, extract two point datasets: locations of capital cities (CI) and locations of countries' centroids (CO). Does a co-location rule "CI → CO" (and vice versa) make sense? In other words, are capital cities usually located centrally in their countries or not?

Ex. 2.9. Discuss the shortcomings of K-anonymity principle, using Figs. 2.14 and 2.15 as the running example. If you were not aware of the quasi-identifier attributes, which tasks would you perform in order to recognize them? How would you proceed in order to identify a "good" value for K?

2.8 Bibliographical Notes

Shekhar and Chawla (2003) is an excellent textbook on spatial databases. The topological relationships discussed in Sect. 2.1 were proposed by Egenhofer and his colleagues back in late '80s and early '90s (Egenhofer 1989; Egenhofer and Herring 1991). The so-called 4- and 9-intersection models are based on the set intersections between objects' interior, boundary (and exterior only for the latter model). Nowadays, they have been adopted by most commercial DBMS in their spatial extensions.

Regarding industrial support to spatial databases, well-known vendors provide it through data types and functions according to the OGC standards. More resources can be found at vendors' web sites, like e.g. in Microsoft (2013), MySQL (2013), Oracle (2013), PostGIS (2013).

The Filter-Refinement methodology for spatial query processing, presented in Sect. 2.2.2, was proposed in Orenstein (1986). R-tree, discussed in Sect. 2.2.2, is the most popular indexing method in spatial databases; the original structure, together with efficient algorithms for index maintenance (insertions, deletions) and querying (point and range queries), was proposed in Guttman (1984). Among dozens of R-tree variations, the most popular include the R+-tree (Sellis et al. 1987) and the R*-tree (Beckmann et al. 1990) while Hilbert R-tree (Kamel and Faloutsos 1994) imposes linear ordering on spatial data (actually, their MBRs) using the Hilbert space-filling curve. Manolopoulos et al. (2005) lists a number of 70 access methods belonging to the R-tree family within a 20 years period (1984–2004). The "ubiquity" of the R-tree is also argued there, with the applications that it has been adopted covering from spatial, image, multimedia, and time-series databases to multi-dimensional (OLAP) analysis. R-tree family seems never ending: indicatively, (Sharifzadeh and Shahabi 2010) recently proposed the VoR-tree, an R-tree variant using Voronoi diagrams for the efficient processing of NN queries.

Speaking about NN queries, the depth-first NN query processing algorithm sketched in Sect. 2.2.2 was originally proposed in Roussopoulos et al. (1995) and then updated by Cheung and Fu (1998), while a best-first alternative was proposed in Hjaltason and Samet (1999). Those works were followed by an extensive list of NN query variations, including *closest-pairs* (Corral et al. 2000), *constrained NN* (Ferhatosmanoglou et al 2001), NN queries under network constraints (Jensen et al. 2003), *all-NN* (Zhang et al. 2004), and *nearest surrounder* queries (Lee et al. 2010). Efficient spatial join query processing using R-trees was proposed in Brinkhoff et al. (1993). Extensions to multi-way joins were studied in Papadias et al. (1999, 2001a), Mamoulis and Papadias (2001). Incremental distance join, where the results are reported one by one ordered by distance, was studied in Hjaltason and Samet (1998). Methods for efficient processing of topological and direction relations in R-tree supported spatial databases were proposed in Papadias et al. (1995), Papadias and Theodoridis (1997).

The performance of R-trees under various types of spatial queries is evaluated, among other works, in Papadopoulos and Manolopoulos (1997), Theodoridis et al.

(2000), Papadias et al. (2001a), Corral et al. (2006). In Manolopoulos et al. (2005), the interested reader ay find plenty of R-tree based algorithms for simple (range, NN, etc.) as well as more complex queries (reverse NN, multi-way spatial join, closest-pair, all-NN, etc.)

The definition about data warehouses that appears in Sect. 2.3 is due to Inmon (1992). Data cubes and the classification of aggregation functions in distributive, algebraic, and holistic were proposed in Gray et al. (1997). Spatial data cubes were proposed in Han et al. (1998). The transition from conventional to spatial data warehouses is discussed in Malinowski and Zimányi (2008). R-trees have been used for OLAP purposes as well: the aggregate R-tree (aR-tree), proposed in Papadias et al. (2001b), processes aggregate queries in spatial data warehouses.

The classification of data mining tasks that appears in Sect. 2.4 is due to Fayyad et al. (1996), also discussed in Andrienko et al. (2008). Tan et al. (2005) is an excellent introductory textbook to data mining, whereas Shekhar et al. (2010) is a recent survey on spatial data mining. Regarding the clustering algorithms mentioned in Sect. 2.4.1, K-means was originally proposed in MacQueen (1967) and finds implementations in almost every data analysis software under different names. DBSCAN and its extension, OPTICS, were proposed in Ester et al. (1996) and Ankerst et al. (1999), respectively. The pseudocode of the DBSCAN algorithm presented in Sect. 2.4.1 and Fig. 2.11 are taken from the respective lemma in Wikipedia.org. Other efficient clustering technique include BIRCH (Zhang et al. 1996), which is based on a pre-clustering scheme to find a first set of clusters and then continues in a hierarchical fashion, the grid-based algorithm STING (Wang et al. 1997), CURE (Guha et al. 1998), which is based on sampling and partitioning, and C^2P (Nanopoulos et al. 2001), which exploits on closest-pairs search in R-trees. Clustering spatial data in the presence of obstacles is discussed in Tung et al. (2001). Sampling in large spatial datasets has been studied in Kollios et al. (2002), Nanopoulos et al. (2006). Spatial co-location pattern mining has been discussed in Shekhar and Huang (2001).

Regarding relational data anonymity, the K-anonymity principle, presented in Sect. 2.5, was introduced in the seminal paper by Sweeney (2002). To overcome K-anonymity shortcomings, several variations have been proposed, including l-diversity (Machanavajjhala et al. 2006) and t-closeness (Li et al. 2007). In particular, a dataset is said to satisfy l-diversity if each equivalence class has at least l 'well-represented' values for the sensitive attribute. Parameter l corresponds to the number of distinct values of a sensitive attribute and defines the degree of protection of the dataset. As the number of l increases, the adversary needs more information in order to extract possible values of sensitive attributes. On the other hand, a dataset is said to satisfy t-closeness if for each equivalence class, the distance between the distribution of a sensitive attribute in the class and the distribution of the attribute in the whole dataset is no more than a threshold t. For further insights into this research area, the reader is referred to the recent survey paper by Fung et al. (2010). Linking data mining with data privacy, PPDM was first discussed in the seminal work by Agrawal and Srikant (2000). The classification of PPDM techniques that is mentioned in Sect. 2.5 is due to Verykios et al. (2004).

References

Agrawal R, Srikant R (2000) Privacy-preserving data mining. In: Proceedings of SIGMOD

Andrienko N, Andrienko G, Pelekis N, Spaccapietra S (2008) Basic concepts of movement data. In: Giannotti F, Pedreschi D (eds) Mobility, data mining and privacy—geographic knowledge discovery. Springer, New York, pp 15–38

Ankerst M, Breunig MM, Kriegel HP, Sander J (1999) OPTICS: ordering points to identify the clustering structure. In: Proceedings of SIGMOD

Beckmann N, Kriegel HP, Schneider R, Seeger B (1990) The R*-tree: an efficient and robust access method for points and rectangles. In: Proceedings of SIGMOD

Brinkhoff T, Kriegel HP, Seeger B (1993) Efficient processing of spatial joins using R-trees. In: Proceedings of SIGMOD

Cheung KL, Fu A (1998) Enhanced nearest neighbor search on the R-tree. SIGMOD Rec 27(3):16–21

Corral A, Manolopoulos Y, Theodoridis Y, Vassilakopoulos M (2000) Closest pair queries in spatial databases. In: Proceedings of SIGMOD

Corral A, Manolopoulos Y, Theodoridis Y, Vassilakopoulos M (2006) Cost models for distance joins queries using R-trees. Data Knowl Eng 57(1):1–36

Egenhofer MJ (1989) A formal definition of binary topological relationships. In: Proceedings of FODO

Egenhofer MJ, Herring J (1991) Categorizing binary topological relations between regions, lines and points in geographic databases. Technical Report, University of Maine

Ester M, Kriegel HP, Sander J, Xu X (1996) A density-based algorithm for discovering clusters in large spatial databases with noise. In: Proceedings of KDD

Fayyad UM, Piatetsky-Shapiro G, Smyth P (1996) From data mining to knowledge discovery: an overview. In: Fayyad UM, Piatetsky-Shapiro G, Smyth P, Uthurusamy R (eds) Advances in knowledge discovery and data mining. MIT Press, Cambridge, MA, pp 1–34

Ferhatosmanoglou H, Stanoi I, Agrawal D, Abbadi A (2001) Constrained nearest neighbor queries. In: Proceedings of SSTD

Fung B, Wang K, Chen R, Yu P (2010) Privacy-preserving data publishing: a survey of recent developments. ACM Comput Surv 42(4):1–55

Gray J, Chaudhuri S, Bosworth A, Layman A, Reichart D, Venkatrao M, Pellow F, Pirahesh H (1997) Data cube: a relational aggregation operator generalizing group-by, cross-tab, and subtotals. Data Min Knowl Disc 1(1):29–53

Guha S, Rastogi R, Shim K (1998) CURE: an efficient clustering algorithm for large databases. In: Proceedings of SIGMOD

Guttman A (1984) R-trees: a dynamic index structure for spatial searching. In: Proceedings of SIGMOD

Han J, Stefanovic N, Koperski K (1998) Selective materialization: an efficient method for spatial data cube construction. In: Proceedings of PAKDD

Hjaltason G, Samet H (1998) Incremental distance join algorithms for spatial databases. In: Proceedings of SIGMOD

Hjaltason G, Samet H (1999) Distance browsing in spatial databases. ACM Trans Database Syst 24(2):265–318

Inmon WH (1992) Building the data warehouse. Wiley, New York

Jensen CS, Kolarvr J, Pedersen TB, Timko I (2003) Nearest neighbor queries in road networks. In: Proceedings of GIS

Kamel I, Faloutsos C (1994) Hilbert R-tree: an improved R-tree using fractals. In: Proceedings of VLDB

Kollios G, Gunopulos D, Koudas N, Berchtold S (2002) Efficient biased sampling for approximate clustering and outlier detection in large datasets. IEEE Trans Knowl Data Eng 15(5):1170–1187

Lee KCK, Lee WC, Leong HV (2010) Nearest surrounder queries. IEEE Trans Knowl Data Eng 22(10):1444–1458

Li N, Li T, Venkatasubramanian S (2007) t-closeness: privacy beyond k-anonymity and l-diversity. In: Proceedings of ICDE

Machanavajjhala A, Gehrke J, Kifer D, Venkitasubramaniam M (2006) l-diversity: privacy beyond k-anonymity. In: Proceedings of ICDE

MacQueen JB (1967) Some methods for classification and analysis of multivariate observations. In: Proceedings of BSMSP

Malinowski E, Zimányi E (2008) Advanced data warehouse design: from conventional to spatial and temporal applications. Springer, New York

Mamoulis N, Papadias D (2001) Multiway spatial joins. ACM Trans Database Syst 26(4):424–475

Manolopoulos Y, Nanopoulos A, Papadopoulos AN, Theodoridis Y (2005) R-trees: theory and applications. Springer, New York

Microsoft (2013) Spatial data (SQL Server). http://msdn.microsoft.com/en-us/library/bb933790.aspx

MySQL (2013) MySQL 5.7 reference manual: 12.18 spatial extensions. http://dev.mysql.com/doc/refman/5.7/en/spatial-extensions.html

Nanopoulos A, Theodoridis Y, Manolopoulos Y (2001) C^2P: clustering based on closest pairs. In: Proceedings of VLDB

Nanopoulos A, Theodoridis Y, Manolopoulos Y (2006) Index-based density biased sampling for clustering applications. Data Knowl Eng 57(1):37–63

Oracle (2013) Oracle spatial developer's guide. http://www.oracle.com/pls/db112/

Orenstein JA (1986) Spatial query processing in an object-oriented database system. In: Proceedings of SIGMOD

Papadias D, Theodoridis Y (1997) Spatial relations, minimum bounding rectangles, and spatial data structures. Int J Geogr Inf Sci 11(2):111–138

Papadias D, Theodoridis Y, Sellis TK, Egenhofer MJ (1995) Topological relations in the world of minimum bounding rectangles: a study with R-trees. In: Proceedings of SIGMOD

Papadias D, Mamoulis N, Theodoridis Y (1999) Processing and optimization of multiway spatial joins using R-trees. In: Proceedings of PODS

Papadias D, Mamoulis N, Theodoridis Y (2001a) Constrained-based processing of multiway spatial joins. Algorithmica 30(2):188–215

Papadias D, Kalnis P, Zhang J, Tao Y (2001b) Efficient OLAP operations in spatial data warehouses. In: Proceedings of SSTD

Papadopoulos A, Manolopoulos Y (1997) Performance of nearest neighbor queries in R-trees. In: Proceedings of ICDT

PostGIS (2013) PostGIS 2.0 Manual. http://postgis.net/docs/manual-2.0/

Roussopoulos N, Kelley S, Vincent F (1995) Nearest neighbor queries. In: Proceedings of SIGMOD

Sellis TK, Roussopoulos N, Faloutsos C (1987) The R+–tree: a dynamic index for multi-dimensional objects. In: Proceedings of VLDB

Sharifzadeh M, Shahabi C (2010) VoR-tree: R-trees with Voronoi diagrams for efficient processing of spatial nearest neighbor queries. In: Proceedings of VLDB

Shekhar S, Chawla S (2003) Spatial databases: a tour. Prentice Hall, Upper Saddle River, NJ

Shekhar S, Huang Y (2001) Discovering spatial co-location patterns. In: Proceedings of SSTD

Shekhar S, Zhang P, Huang Y (2010) Spatial data mining. In: Maimon O, Rokach L (eds) Data mining and knowledge discovery handbook, 2/e. Springer, New York

Sweeney L (2002) K-anonymity: a model for protecting privacy. Int J Uncertain Fuzziness Knowl Based Syst 10(5):557–570

Tan PN, Steinbach M, Kumar V (2005) Introduction to data mining. Addison-Wesley, Boston

Theodoridis Y, Stefanakis E, Sellis TK (2000) Efficient cost models for spatial queries using R-trees. IEEE Trans Knowl Data Eng 12(1):19–32

Tung AKH, Hou J, Han J (2001) Spatial clustering in the presence of obstacles. In: Proceedings of ICDE

Verykios VS, Bertino E, Fovino IN, Parasiliti Provenza L, Saygin Y, Theodoridis Y (2004) State-of-the-art in privacy preserving data mining. SIGMOD Rec 33(1):50–57

Wang W, Yang J, Muntz R (1997) STING: a statistical information grid approach to spatial data mining. In: Proceedings of VLDB

Zhang T, Ramakrishnan R, Linvy M (1996) BIRCH: an efficient data clustering method for very large databases. In: Proceedings of SIGMOD

Zhang J, Mamoulis N, Papadias D, Tao Y (2004) All-nearest-neighbors queries in spatial databases. In: Proceedings of SSDBM

Part II
Mobility Data Management

It is a very sad thing that nowadays there is so little useless information.

Oscar Wilde

Chapter 3
Modeling and Acquiring Mobility Data

The vast spread of GPS equipped mobile devices, such as smart phones, GPS navigation devices etc., combined with the development of appropriate techniques for storing, processing, querying, and mining such kind of data, has resulted in the production of huge amounts of location-aware information. However, getting this kind of raw data into a meaningful form in terms of mobility is not a straightforward task at all. During this transformation procedure, many aspects arise, such as the treatment of inaccurate or noisy information, the identification of trajectories as sequences of sampled positions, the reduction of the size of the datasets in order to deal with the storage challenges that may appear, etc. Furthermore, in order to be able to evaluate the performance of spatiotemporal algorithms and data structures, enormous amounts of real-world mobility data are required, which cannot be easily found available out there. To bridge this gap, many generators of moving object trajectories have been developed. This chapter provides a review on the above-mentioned research field and is organized as follows. After a necessary discussion on mobility data modeling (the concepts of time-stamped locations, trajectories, etc.), we refresh our discussion on collecting raw mobility data through GPS devices and handle aspects that arise, such as noise and inaccuracy. Then, we introduce the trajectory reconstruction problem and the most popular techniques dealing with this issue (trajectory identification, map-matching, compression). We conclude by familiarizing the reader with the notion of mobility data generators along with a discussion on several developments on that field, for movement either in free space or under network constraints.

3.1 Modeling Mobility Data

Mobility data involves the location of moving objects, which evolves with time. Locations can be point or regional (consider, for instance, moving cars vs. moving hurricanes), hence producing *moving points* or *moving regions*, respectively. In this book, the focus in on the former type of movement, so we will not further discuss

N. Pelekis and Y. Theodoridis, *Mobility Data Management and Exploration*,
DOI 10.1007/978-1-4939-0392-4_3, © Springer Science+Business Media New York 2014

moving regions and related applications. This is because the majority of real-world applications abstract objects as points, even if they have an area (consider e.g. the footprint of a car on the road), since this is usually of no interest for the application.

Theoretically speaking, the movement of an object is defined as a mapping from temporal $I \subseteq \mathbb{R}$ to geographical space $S \subseteq \mathbb{R}^2$:

$$I \subseteq \mathbb{R} \rightarrow S \subseteq \mathbb{R}^2 : t \rightarrow l(t)$$

where $l(t)$ is the location where the object was found at time t. As such, the (actual) *moving object trajectory* is defined as:

$$T_{act} = \{(t, l(t)) | t \in I\} \subset \mathbb{R}^2 \times \mathbb{R}$$

Of course, this represents a continuous curve in reality, but it is commonly accepted that reality cannot be fully implemented in computer systems! Therefore, while the actual trajectory T_{act} of a moving object is a 3-dimensional space *curve*, in fact, in computer systems it is represented in a finite version, the most popular of which being a sequence of time-stamped locations:

$$T = \{< p_1, t_1 >, < p_2, t_2 >, \ldots, < p_n, t_n >\},$$

where $p_i \in \mathbb{R}^2, t_i \in \mathbb{R}, 1 \leq i \leq n$, and $t_1 < t_2 < \ldots < t_n$.

Mapping from T to T_{act}, is not unique since T can be considered as lossy compression of T_{act}, and by definition discards information. Still, the fact that our knowledge about the object's movement is discrete (instead of continuous) raises the following natural question: *where was the object located at a time between two sampled timestamps?* Whatever is the answer, it includes an inherent uncertainty depending on the sampling rate and movement parameters, with speed being the most critical. Nevertheless, several interpolation techniques for approximating T_{act} have been proposed in the literature, with linear interpolation being the most popular one, because of its simple definition and easy implementation.

According to *linear interpolation*, illustrated in Fig. 3.1, time-stamped locations <p_i, t_i> are connected with straight lines in 3-dimensional space, and the location an object was found at a (non-recorded) timestamp $t \in (t_i, t_{i+1})$, where t_i and t_{i+1} are two consecutive recorded timestamps, is calculated as follows:

$$p(t) = \left(x_i + \frac{t - t_i}{t_{i+1} - t_i}(x_{i+1} - x_i), y_i + \frac{t - t_i}{t_{i+1} - t_i}(y_{i+1} - y_i) \right)$$

The basic assumption followed by this formula is that the velocity vector (i.e. speed and direction) remains constant during time interval $[t_i, t_{i+1})$, which (unfortunately) results in discontinuous evolution of those movement parameters, exactly at the recorded timestamps t_i.

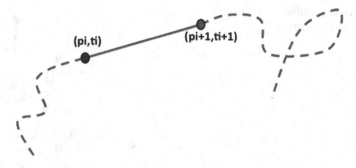

Fig. 3.1 Linear interpolation between sampled locations

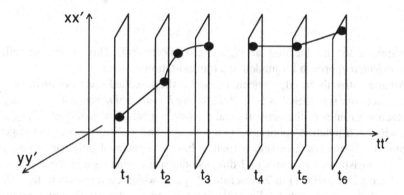

Fig. 3.2 The sliced representation for the definition of moving objects

The result of the above discussion is that the "trajectory" of a moving object can be modeled according to the concept of *sliced representation*: this model decomposes the temporal development of a value into fragments called *slices*, such that within a slice this development can be described by some kind of *simple* function. Sliced representation is illustrated in Fig. 3.2, where dots represent the sampled positions and lines in between represent alternative interpolation models (linear vs. arc interpolation). Unknown type of motion may also be found in a trajectory, as is the case in the interval $[t_3, t_4)$ of Fig. 3.2.

The above discussion assumes movement in the (unconstrained) Euclidean space. However, in several applications, moving objects are assumed to move along transportation networks. Formally, a transportation network is modeled as $G=(V, E)$, where V is a set of m vertices $\{v_1, v_2, ...,v_m\}$, representing e.g. road junctions, and $E \subseteq V \times V$ is a set of n edges $\{e_1, e_2, ..., e_n\}$, with each edge connecting two vertices, v_{from} and v_{to}, and representing e.g. routes between road junctions. With respect to the geographical information involved, a vertex v has an associated point $p=(x, y)$ in 2-dimensional space whereas an edge e has an associated polyline $l=<p_1, p_2, ...,p_k>$ consisting of k points in 2-dimensional space. The location of a moving object on an edge e (v_{from}, v_{to}) is represented by a real number in $[0..1]$, where value 0 (1)

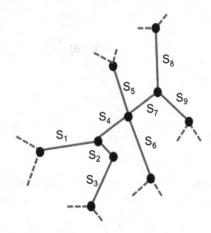

corresponds to the location at v_{from} (v_{to}, respectively). This is the so-called *edge-oriented* approach for modeling a transportation network.

An alternative to the edge-oriented model is the so-called *route-oriented* model. Here, a network is modeled as $G' = (R, J)$, where R is a set of n' routes $\{r_1, r_2, ...,r_n'\}$, representing routes of the network, and $J \subseteq R \times R$ is a set of m' junctions $\{j_1, j_2, ..., j_{m'}\}$, with each junction connecting two routes, r_i and r_j. With respect to the geographical information involved, a route r has an associated polyline $l = <p_1, p_2, ...,p_k'>$ consisting of k' points in 2-dimensional space whereas a junction j has an associated point $p = (x, y)$ in 2-dimensional space, which corresponds to the intersection of the associated polylines of routes r_i and r_j. The location of a moving object on a route r is represented by a real number in $[0..1]$, where value 0 (1) corresponds to location p_1 (p_k', respectively) of the associated polyline of r.

Finally, the simplest version of a network model is the so-called *segment-oriented* model. In this model, every single segment of the network is an edge in the graph connecting two points. Figure 3.3 illustrates an example of transportation network consisting of 9 segments (S_1, ...,S_9). According to the edge-oriented model, this networks consists of 8 edges: $<S_1>$, $<S_2, S_3>$, $<S_4>$, $<S_5>$, $<S_6>$, $<S_7>$, $<S_8>$, and $<S_9>$. On the other hand, according to the route-oriented model, this network might consist of 4 routes: $<S_1, S_4, S_7>$, $<S_2, S_3>$, $<S_5, S_6>$, and $<S_8, S_9>$.

3.2   Acquiring Trajectories from Raw Data

As discussed in Chap. 1, GPS is a satellite-based global positioning system, which broadcasts precise timing signals to GPS receivers, allowing them to accurately determine their location in 3-dimensional space (longitude, latitude, and altitude).

An important characteristic of GPS recorded data is the *sampling rate*, i.e. "how often" we get a position reading. Sampling rate could be either constant or variable. Constant sampling rate means that we get readings in regular intervals, for instance, every 10 s. Let us consider a driver in a highway, keeping stable velocity

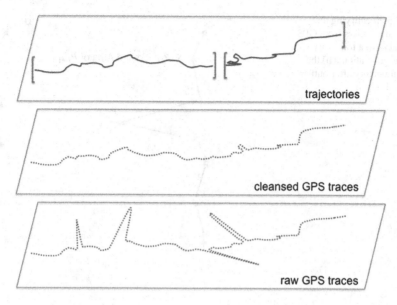

Fig. 3.4 From raw to cleansed GPS data, and then to (segmented) trajectories of moving objects

vector for minutes. By sampling her position, for instance, every 10 s, we make unnecessary use of the storage resources of the GPS device. In an effort of trying to make the tracking devices more efficient, in terms of energy consumption and generally resources usage, several techniques have been developed. Targeting at having low variable sampling rate and at the same time minimizing the loss of information that is implied, these techniques take into account speed, acceleration, heading, etc. In the above example of the driver in a highway, there is no useful-ness in taking samples during the "constant movement" period. On the other hand, at the moment that the car changes significantly its speed and/or heading we may take a sample, and the movement during the "silent" period may be inferred by performing linear interpolation between the previous and the current sample.

Due to sampling and other factors, *uncertainty* of mobility data appears. In order to record the exact movement of an object, theoretically speaking, its position at every single moment would be needed, which conflicts to the discrete nature of sampling as already discussed. In addition to the effect of sampling, another source of uncertainty is the measurement error of GPS devices or even noise. For example, current GPS technology determines an object's position with an accuracy of a few meters. Also, in case of a system malfunction, we receive a noisy reading that is inaccurate enough or even unacceptable, taking Physics laws into account.

In general, in order to reconstruct meaningful trajectories of moving objects from raw GPS data, there are two main processing steps: (a) *data cleansing*, which cleans raw data by smoothing values, removing outliers, etc., and (b) *trajectory identification*, which identifies trajectories from cleansed data (i.e. approximates the actual trajectory, also identifies starting and ending points of trajectories). The tra-jectory reconstruction problem is illustrated in Fig. 3.4 (note that the temporal

Fig. 3.5 Identifying and
removing outliers from GPS
raw data: (**a**) a trajectory with
an outlier location; (**b**) the
same trajectory after outlier
removal

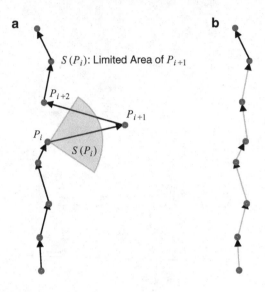

information is omitted in the figure): the input consists of a sequence of, perhaps
dirty, time-stamped locations (bottom), it is cleansed (middle), and the output consists of well-formed and separated trajectories (top).

The second step could be extended, if necessary, to include trajectory simplifica-
tion/compression (to address storage issues) and map-matching (to exploit the
knowledge that movement is along a transportation network). This two-step procedure is discussed in the following subsections.

3.2.1 GPS Data Cleansing

As already mentioned, raw mobility data in real life is often noisy and erroneous.
Consequently, it needs to be cleansed in order to provide us with more precise information about an object's movement. The primary task of data cleansing is the minimization or the total removal, if possible, of GPS errors. GPS errors, as already mentioned,
are incorrect measurements of the spatial position and may appear due to several reasons: insufficient number of satellites in view, satellite orbit, receiver deficiency, etc.
The resulting error is on the spatial information, as the temporal dimension is assumed
to be accurate because of the high calibration clocks that the satellites utilize.

Erroneous recordings may be classified as *noise*, which are recordings that are
impossible to happen in reality due to Physics laws, or *random errors*, which are
small (a few meters long) deviations from true values; the former should be removed
while the latter may be smoothed. A widely acceptable method for removing noise
is the design and implementation of automated 'filters'. For example, noisy positions
could be filtered out by taking into account the maximum allowed speed of a moving
object, V_{max}, in order for a position not to be considered as noise with respect to the
previous one. The basic concept is to define an area $S(p_i)$ for each point p_i under
study, and inspect whether p_i lies inside area $S(p_i)$ or not. For example, in Fig. 3.5, in

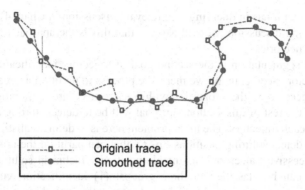

Fig. 3.6 GPS data cleansing by smoothing random errors

order to compute area $S(p_i)$, the previous point p_{i-1}, the temporal duration $(t_i - t_{i-1})$ and V_{max} are taken into consideration. If p_i is found to lie outside area $S(p_i)$ it is considered as noise and needs to be removed.

On the other hand, removal of random errors is not necessary since smoothing is feasible instead. An approach for smoothing could be based on the least squares spline approximation that targets to minimize overall error terms or a kernel-based method, which adjusts the probability of occurrences in the data stream to modify outliers. An example of GPS data cleansing by smoothing random errors appears in Fig. 3.6.

Nevertheless, raw mobility data is nothing more than time-stamped locations. In order to get them into a meaningful form in terms of mobility, these positions have to be converted into meaningful spatiotemporal trajectories.

3.2.2   Trajectory Identification

Trajectory *identification* (the second step of trajectory reconstruction) targets at interpolating successive positions $<p_i, t_i>$ in order for the actual trajectory of the moving object to be approximated (the first goal of trajectory identification) and segmenting the sequences of positions in homogeneous partitions (the second goal of trajectory identification). This process is not as straightforward as it may sound. If we consider that raw mobility data (positions) are received by a trajectory database server in bulk loads, a procedure is required that determines whether the newly arrived data is to be attached to an existing trajectory (and, in such case, how to be linked to the existing "tail" of the trajectory), or initiate a new trajectory.

The first goal of trajectory identification involves successive positions and interpolates them in order to simulate the *continuous* nature of the actual movement of an object. Several interpolation techniques have been proposed in the literature (linear interpolation, Bezier curves, etc.). Linear interpolation, already discussed in Fig. 3.1, has been shown to be the most popular, especially because of its simple definition and easy implementation. On the other hand, it appears to have a few deficiencies, such as the discontinuous speed and direction assumption it makes as well as the effect it has in querying (see relevant discussion in Chap. 4), but in the remainder of the discussion, we will assume that this is the approach followed to reconstruct trajectories.

Trajectory segmentation is the second goal of trajectory identification. In general, by trajectory segmentation we mean the process to partition a stream of positions into pieces so as the properties within each piece are as homogeneous as possible. For the rest of this section, our goal will be to detect starting and ending positions for each trajectory. The main problem here is to define realistic identification rules, to detect splitting positions $<p_i, t_i>$ that can partition the non-stop GPS feed into successive trajectories at suitable positions. Technical solutions for this sub-problem can be classified in three groups: (1) identification via raw gap, (2) identification via prior knowledge, and (3) correlation-based identification.

Identification via Raw Gap: A spatiotemporal sequence of positions $<p_i, t_i>$ can be partitioned into separate trajectories with respect to the spatiotemporal gaps that appear in the raw sequence (note that we have already removed the noise in a previous step). At least two parameters should be taken into account, either in conjunction or not: the temporal and the spatial gap, defined as follows:

– Spatial gap: the maximum distance, S_{gap}, between two consecutive time-stamped positions of the same object in order to be considered that they belong to the same trajectory.
– Temporal gap: the maximum allowed time interval, T_{gap}, between two consecutive time-stamped positions of the same object in order to be considered that they belong to the same trajectory.

Having set S_{gap} and T_{gap}, as soon as we find a pair $<p_{i-1}, t_{i-1}>$ and $<p_i, t_i>$ of sampled positions, such as their spatial and/or temporal distance exceeds S_{gap} and T_{gap}, respectively, then we identify $<p_{i-1}, t_{i-1}>$ as the ending point of the existing trajectory and $<p_i, t_i>$ as the starting point of a new trajectory of that moving object. Obviously, setting these parameters is application-dependent; e.g. $S_{gap} = 1$ km and $T_{gap} = 1$ h may sound reasonable for some applications while for others it may not.

Identification via Prior Knowledge: Apart from the usage of raw gap, some prior knowledge could be used in the service of trajectory identification. Such prior knowledge might include predefined time intervals or predefined spatial extents. Predefined time intervals could divide the GPS sequence into a number of subsequences included in certain time intervals. For example, if, for a specific application, there is prior knowledge that during weekdays workers usually move in the morning (around 8 am to 9 am) and, then, in the afternoon (around 4 pm to 5 pm), this knowledge could be used for trajectory identification purposes. Similarly, predefined spatial extents could divide the GPS sequence into a number of subsequences according to a spatial criterion. Such a criterion could be based on a fixed distance threshold (e.g. every 1 km a trajectory ends and a new one is initiated) or movement along predefined regions (e.g. regions representing ports in a vessel traffic application) or points (e.g. bus terminals in a bus routing application).

Correlation-based Identification: In a different treatment of the problem, we can apply correlation-based techniques to discover spatiotemporal trajectories. Here, no prior knowledge is needed in advance. The splitting up of the cleansed trajectory is purely reliant on the correlations between positions $<p_i, t_i>$. For example, conventional time series segmentation methods could be applied for the discovery of division points between two successive spatiotemporal trajectories.

3.3 Trajectory Reconstruction and Simplification

Data smoothing techniques discussed in Sect. 3.2, have been designed to work upon objects moving in free space. However, there exist several real-world applications that consider objects restricted to move along a given transportation network, usually a road network. As already explained, due to errors, noise, etc., although in reality an object (private car, taxi, bus) moves along the network, the recorded positions may be found outside it. So, an extra pre-processing step that matches the trajectories to the network is needed. This process is commonly called *map-matching*. Technically, map-matching refers to the process of identifying the correct network segment (e.g. road segment), on which the object of interest is moving and its exact position on that segment.

3.3.1 Trajectory Reconstruction via Map-Matching

Having defined a network as a directed graph $G = (V, E)$, following e.g. the edge-oriented model discussed in Sect. 3.1, a network-constrained trajectory can be properly determined: for any given position $<p_i, t_i>$ of a trajectory, p_i should lie on an edge e_j. In other words, if $<p_i, t_i>$ does not match the network, it needs to be replaced by $<p_i', t_i'>$, such that $p_i' \in e_j$ and t_i' is similar (not necessarily equal) to t_i.

Map-matching techniques can be classified into online (when positions arrive at real time) vs. offline (when all positions are available beforehand). *Online* map-matching algorithms are applied to real-time applications, such as navigation systems. Matching is performed as soon as current data is being received, consequently it can take into account only current and past data. On the other hand, *offline* algorithms are applied to post-processing applications, which can also take advantage of "future" positions, a fact that is quite helpful for the algorithm to select the correct matching. Both groups can be further categorized as geometric, topological, probabilistic, and advanced hybrid methods.

Geometric map-matching. Geometric map-matching algorithms base the match on the geometry of the network segments, choosing the one that is closest to the position of the moving object. They are divided into three types: *point-to-point* (e.g. Euclidean distance), *point-to-curve* (e.g. perpendicular distance), and *curve-to* *curve* (e.g. Fréchet distance). The first type matches the position of the moving object to the closest node belonging to the network; the second type prefers the closest network segment. The line segment that provides the minimum distance is chosen as the one on which the moving object is actually travelling; the third type is based on the point-to-point match where it selects a set of candidates and then the final curve chosen is the one closest to the curve composed by the current points being matched. The route with the smallest distance from the initial trajectory is taken as the map-matched trajectory. To make the above discussion clear, Fig. 3.7 provides three respective examples.

Topological map-matching. Topological algorithms utilize the geometry as much as the connectivity and the adjacency of the graph representing the network. Nevertheless, they generally ignore additional features of movement, such as speed and direction, and might be sensitive to outliers as well. These algorithms can be divided into two steps: the first one is the initial matching process, where the algorithm chooses the most suitable link from the closest to the initial points; during the second step, the algorithm continues matching the points considering the network topology.

Figure 3.8a illustrates an example of topological map-matching, based on a state-of-the-art methodology. Given that point P_{i-1} has already been matched to edge c_3, the candidate edges are the incident edges "exiting" the last matched edge (including also the matched edge itself); for P_i they are c_1, c_2 and c_3. Two similarity measures are employed for the assessment and selection among the candidate edges, applying distance and orientation criteria, respectively. The higher the weighted sum of these measures is, the fittest the edge is to be matched with the point under study. In the example of Fig. 3.8a, c_1 is the selected edge for P_i. On the other hand, if the projection of the current point on the candidate edges does not lie on any of these edges, the algorithm does not proceed to the next point. Instead, the nearest edge of the candidates is set as part of the trajectory and then the next set of candidate edges is evaluated.

In addition, a "look ahead" approach may improve the quality of the above decision. In detail, for every candidate edge the score of the best candidate among its adjacent accessible edges is calculated. The main purpose is the exploration of alternative branches instead of edges in order to lead to a more spherical decision making process. Of course, in the end, only one choice appears in the matching result. More specifically, Fig. 3.8b illustrates an example with look-ahead of two edges. Having e_2 as the candidate edge to be matched, $e_{2,1}$ is the best candidate for matching point P_{i+1} assuming that P_i is matched to edge e_2. Considering e_1 as the candidate edge, it is obvious that is also the fittest candidate for point P_{i+1}, given the fact that P_i is matched to e_1. The overall scores for matching point P_i to edges e_1, e_2 and e_3 are calculated as the sum of scores for every best subpath as

$$s\left(p_i,e_j\right) = \sum_{k,l=0}^{depth} s\left(p_{i+l},e_{j,k}\right)$$

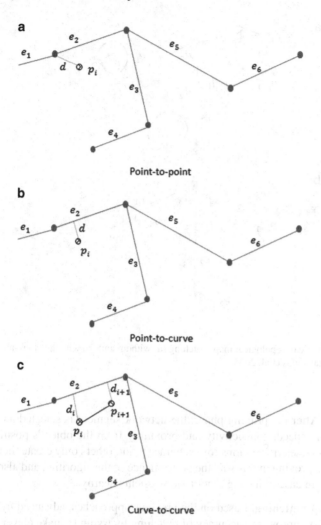

Fig. 3.7 Examples of geometric map-matching, following (**a**) point-to-point, (**b**) point-to-curve, and (**c**) curve-to-curve approach

Probabilistic map-matching. Probabilistic algorithms take advantage of *error region*, usually an ellipse or a rectangle, to match the given point. The error region is the potential area of movement of an object, given a maximum speed. More specifically, this region is utilized for pruning the search space and selecting only the network segments that lie in the potential area of movement of the object

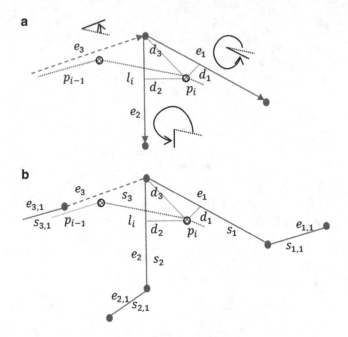

Fig. 3.8 Example of topological map-matching (**a**) without and (**b**) with the "look-ahead" option. Source: (Brakatsoulas et al. 2005)

of interest. After the pruning phase, the network segment is selected according to the direction, speed, connectivity and proximity from the object's position. Some algorithms create error regions for each trace point, others only create them close to junctions, improving in this way the performance of the algorithm and also avoiding mismatches in case of having other road segments nearby.

Hybrid map-matching. Based on the previous approaches, advanced hybrid algorithms aim to improve the accuracy of matching by trying to make clever combinations. Aside from location, these algorithms are often supplemented with extra information such as speed, direction, network connectivity, quality of the input data or even using correction errors from third party systems (e.g. Assisted GPS, see the related discussion in Chap. 1).

Of course, whichever is the algorithm, the accuracy of the map-matching process highly depends on map resolution and completeness: the higher the resolution, the more accurate the matching will be. In most cases, the algorithms assume that the map network is complete and accurate. This is often a wrong assumption, and algorithms might appear to have unexpected behaviour when, for instance, there could be found no network segment nearby to match. From this brief remark, it turns out that apart from the goal of mapping a moving object to a road network, map-matching could also work reversely: moving object trajectories that are originally unmatched to the road network could be used to improve the quality of the network itself, especially if this information comes from extremely valuable, though not necessarily accurate open sources. Although very interesting, this perspective will not be discussed further.

3.3.2 Trajectory Simplification via Data Compression

Trajectory data is typically produced continuously with a high sampling rate. This kind of enormous amounts of trajectory data can sooner or later lead to storage, transmission, computation, and display challenges. As a result, trajectory simplification through data compression turns out to be a tool for supporting the scalability of mobility data intensive applications.

Generally, trajectory compression algorithms can be categorized into four groups: top-down, bottom-up, sliding window, and opening window. The *top-down* methods are based on recursively splitting the sequence of positions and only keeping the representative positions in each sub-sequence. The *bottom-up* algorithms start from the finest possible representation, and merge the successive points until the ending conditions are met. In the *sliding window* approach, data are compressed in a fixed window size; whilst in *open window* methods, a dynamic and flexible data segment size is used.

An example of a top-down method is the Synchronous Euclidean Distance (SED), which is based upon the popular Douglas-Peucker algorithm for line simplification on the plane. Before we present SED, let us briefly discuss Douglas-Peucker. The main functionality of Douglas-Peucker algorithm is the reduction of the number of points in a curve that is approximated by a significantly smaller series of points. Given a curve consisting of a set of line segments, the goal of the algorithm is to discover an analogous curve consisting of less number of points, selected among the original ones. Here, "similarity" is defined upon the maximum distance between the original and the simplified curve. Practically, this method computes the perpendicular distance of each internal point from the line connecting the first and the last point of the original curve (line segment AB in Fig. 3.9a) and discovers the point with the maximum perpendicular distance (point C). Then, it recursively performs the same procedure with the curve now considered to be the polyline ABC produced due to the aforementioned "breaking point" (point C), and so on. The ending condition of this algorithm is met when the distance of all the remaining points from the currently examined line is less than a given threshold (e.g. all points between C and B with respect to line segment BC) or there are no points left to examine. In the first case the algorithm ends and returns this line segment as part of the new—compressed—polyline.

Returning to the spatiotemporal domain, it is obvious that Douglas-Peucker does not take the time dimension into consideration since it is a (geometrical) line simplification algorithm. Its extension in spatiotemporal domain, SED, replaces the perpendicular distance used in Douglas-Peucker for choosing the "breaking points" with the so-called *Synchronous Euclidean Distance*, which is the distance among the currently studied point (P_i in Fig. 3.9b and the point of the line segment P_sP_e

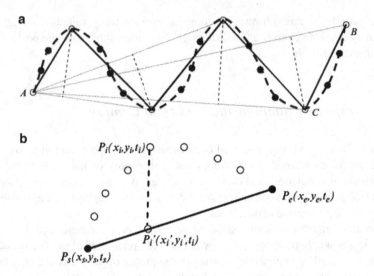

Fig. 3.9 (a) Douglas-Peucker line simplification; (b) synchronous Euclidean distance (SED)

where the moving object would lie, supposed it was moving on this line, at time instance t_i determined by the point under examination (P_i' in Fig. 3.9b). The goal here is to recursively compute those distances for each sequence's in-between points, and compare them with a given threshold ε to choose the "breaking points" (in other words, to decide whether a point can be removed or not).

In the literature, there also exist approaches that combine compression and map-matching. The main effort there is, given a transportation network $G=(V, E)$ and a (unmatched with respect to the network) trajectory T consisting of positions $<p_i, t_i>$, to find a both network-constrained and compressed counterpart of T, called T'. To deal with this problem, there are at least two approaches that can be followed: (a) compress the original trajectory T and afterwards perform map-matching to find T' or (b) map-match the original trajectory T, then compress it and, eventually, again map-match it to result to find T'. Since the final result (T') is not necessarily the trajectory actually followed by the moving object (due to compression, T' may significantly differ from the map-matched version of T), these methods may find applications in a framework where approximate (sketch) rather than the actual trajectories are required, e.g. for high-level traffic analysis.

3.4 Trajectory Data Generators

One of the most important requirements for spatiotemporal algorithms and data structures is the exploitation of understandable and easy to use generators of synthetic trajectory data for performance evaluation purposes. Using these methods is extensive, as they facilitate the study of MODs. The generators created so far are divided into two categories: (1) those supporting movement in free space and (2) those supporting network-constrained movement. Some of them have limited functionality (e.g. limited ability to choose direction, speed, etc.), while others provide more features. Another categorization that could be made is *microscopic* vs. *macroscopic* data generators. Microscopic generators deal with single vehicles; the majority of them have been developed to quantify the benefits of Intelligent Transportation Systems (ITS) and in terms of scale the variation ranges from a few vehicles and nodes (i.e. road intersections) to a large number (about 200 nodes and many thousands of vehicles). The models produced by the microscopic generators are mostly used to evaluate traffic efficiency regarding speed and travel time. They basically focus on simulating traffic signal regulation, route control and traffic condition estimation. On the other hand, macroscopic generators deal with the traffic flow (not with single vehicles) at a higher abstraction level than the microscopic generators. They are employed to evaluate traffic flow as a whole without consideration of the characteristics and features of individual vehicles in the traffic. In the remainder of this section, we will present, in chronological order, some of the most popular data generators (classified as macroscopic, according to the above discussion).

3.4.1 Generating Trajectories in Free Space

GSTD (Generating Spatio-Temporal Data) was one of the first moving object generators to appear in the literature. It generates time stamped locations of moving objects (points or rectangles) in free space. As simulation time passes, objects change their location according to a number of parameters (speed, heading, etc.) and size (in case of non-point objects). The original algorithm found several improvements later, allowing more options to users: agility of generated objects, consideration of obstacles, forced passing from pre-determined points, etc. Figure 3.10 illustrates four examples of running *GSTD* simulating different movement behaviours: slow vs. fast moving objects, biased movement towards one or more directions (hence, clusters of moving objects could be also generated).

Inspired by *GSTD*, *CENTRE* (Cellular Network Trajectories Reconstruction Environment) examines movements of objects in mobile networks, using data from cell phone providers and simulating larger numbers of objects. Its most interesting characteristic is that groups with different movement behavior (direction, speed, agility) and sensitivity to obstacles can be defined. For example, as illustrated in Fig. 3.11, cars can be obstructed by rectangular obstacles (e.g. building blocks) while pedestrians not.

3.4.2 Generating Network-Constrained Trajectories

Brinkhoff was the earliest attempt to generate moving objects over a network. It is an open-source generator, with graphical interface for objects moving over a network.

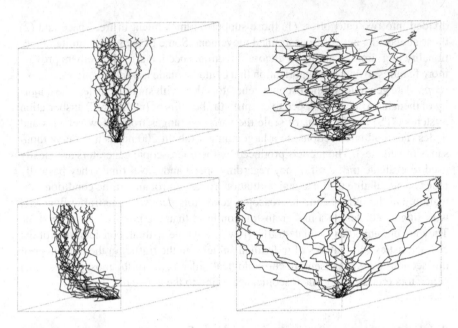

Fig. 3.10 Screenshots from GSTD synthetic data generator in free space (clockwise, starting from upper-left): slow moving objects, fast moving objects, directed movement towards the four corners of the horizon, directed movement towards south. Source: (Pfoser and Theodoridis 2003)

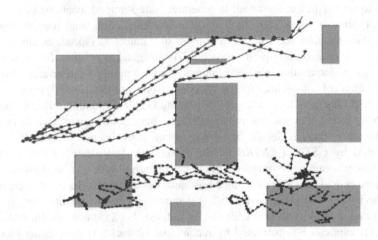

Fig. 3.11 Screenshot from CENTRE synthetic data generator in free space with obstacles

During the simulation, new objects are created in random locations on the grid and disappear when they reach their destination. These objects are separated during their creation in groups (can be displayed in different colors) to highlight different properties they may have. The speed and the path followed by each object depend on the group to which it belongs, but also from the rest of the population.

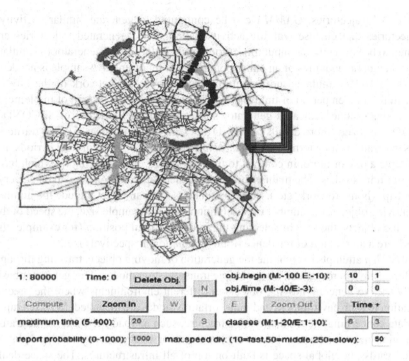

1 : 80000	Time: 0	Delete Obj.		obj./begin (M:-100 E:-10):	10	1
			N	obj./time (M:-40/E:-3):	0	0
Compute	Zoom In	W	E	Zoom Out	Time +	
maximum time (5-400):	20	S	classes (M:1-20/E:1-10):	8	3	
report probability (0-1000):	1000	max.speed div. (10=fast,50=middle,250=slow):			50	

Fig. 3.12 Screenshot from Brinkhoff network-constrained synthetic data generator

This is because each edge of the network has a maximum capacity, which, if exceeded, affects the speed of moving objects in it. The basic functions, such as the number of moving objects or timing simulation of the generator can be regulated by a limited set of parameters that are easily changed, not necessarily from experienced users. A screenshot of *Brinkhoff* generator appears in Fig. 3.12.

Another example of moving object simulator over a network is *SUMO* (Simulation of Urban Mobility), an open source generator that simulates the movements of private and public transport. In *SUMO* simulations, vehicle movements are represented in a network that does not allow conflicts, streets are two-way, etc., thus completing the picture of a real network. An additional feature of *SUMO* generator is that the behavior of moving objects generated is based on the proper interaction between them. So the movement of each object that is created depends on the motion of other objects in the network design.

GAMMA framework represents moving object behavior as spatiotemporal trajectories and recommends two metrics to evaluate trajectory data sets. Here, the generation process is considered as an optimization problem and is solved by the utilization of a genetic algorithm. In fact, this framework is a *generator-by-example*, which means that it receives as input a set of trajectories and tries to generate both cellular network trajectories and symbolic trajectories that (a) represent mobility patterns similar to those present in a set of real–life sample trajectories given as input and (b) conform to real–life constraints and heuristics. Hence, based on these sample "activity" trajectories, *GAMMA* can be configured to generate similar "activity" trajectories that enclose real–life activity patterns. The generated trajectories are aimed to be similar to the input trajectories. So in order for the generator to simulate spatiotemporal activities of an entire population, a representative sample is needed.

BerlinMOD simulates movements of objects over the network of the city of Berlin for a given period of time and capture their positions. Instead of implementing a stand alone dedicated generator software, the infrastructure of SECONDO MOD is utilized (more details on SECONDO are found in Chap. 5). The traffic in this generator is long term, i.e. simulates movements of objects for long periods, for example a day or more, in contrast to the majority of other generators, which have shorter time strokes. The main effort of this generator is to model a person's everyday trips (home to work, etc.). Some parameters that can be set include the number of moving objects, the number of observation days, the sample size, the speed of the moving objects, the way of selecting the initial and final position (for example, the work are a and region of residence of the population, respectively), etc.

MWGen attempts to combine the generation of moving objects traveling through different environments and with multiple transportation modes. This generator basically tries to generate moving objects in different environments where the precise location in each environment and transportation modes are managed. MWGen input is data concerning the available infrastructure, such as road network, public transportation network (metro, buses, etc.) and indoor environment (i.e. floor plans). Afterwards, the global space is built on top of all infrastructures. The space deals with the representation for each infrastructure, loads infrastructure objects if they are needed for query processing, and controls the interaction or linking parts along the different infrastructures. Finally, trip planning is performed on the unified space where the start and end positions can be in any given infrastructure. A trip can cover one or several environments, where in the latter case the moving object uses multiple transportation modes.

Finally, *Hermoupolis* follows an approach that combines a moving object data generator and an activity (in terms of diaries) generator. Actually, *Hermoupolis* produces both network constrained trajectories and *semantically enriched* trajectories (the semantic aspects of moving object trajectories will be discussed in depth in Chap. 9). The input of this generator is socio-demographic data, land use data, points of interest (POI), the underlying network and some example semantically annotated trajectories to be used as the seeds of the generating procedure (as such, *Hermoupolis* is also a *generator-by-example*). A screenshot of *Hermoupolis* generator appears in Fig. 3.13.

In detail, the generating process starts with building the profiles of moving objects through applying clustering on the seed trajectories. The goal of this step is to discover the profiles of the moving objects (persons) and a representative semantically annotated trajectory for each profile. The next step takes as input these representatives along with the land use data and the POIs, and generates two outputs, network-constrained trajectories and their semantically annotated counterpart, by utilizing *Brinkhoff* data generator presented earlier. It is worth to be mentioned that both outputs are synchronized, i.e. GPS-like positions correspond to semantic annotations (home, work, driving, walking, etc.) and vice versa.

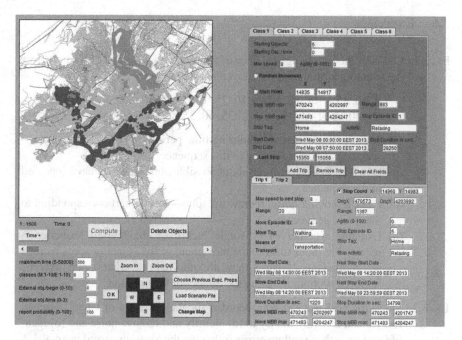

Fig. 3.13 Screenshot from Hermoupolis network-constrained synthetic data generator

3.5 Summary

This chapter focused on GPS-like positioning data and discussed several modeling as well as preparatory issues that arise before a MOD is ready to productive work. The most critical problem we face is the *trajectory reconstruction*, in which the original, inaccurate and perhaps noisy, data fed by the data collection step have to be converted into reasonable spatiotemporal trajectories. This includes two steps: data cleansing and trajectory identification. During the first step, the goal is the minimization of the effect of noise or random errors produced by GPS. This is achieved by applying smoothing filtering and map matching techniques. The output of data cleansing becomes the input for the trajectory identification, which can be addressed by using simple spatial and/or temporal gaps or even more advanced methods.

Other important preparatory operations include trajectory map-matching and compression (simplification). In the former, trajectories are mapped to underlying transportation networks, assuming that movement is network-constrained, while in the latter, trajectories are simplified through compression in order to deal with the storage, transmission, computation, and display challenges that may appear.

Finally, this chapter surveyed trajectory data generators. In order to be able to proceed to performance evaluation for spatiotemporal algorithms and data structures, enormous amounts of data are required. Data generators of moving objects address this issue since real scalable mobility datasets cannot be easily found for public use.

3.6   Exercises

Ex. 3.1. Consider the following sequence of ten recorded positions of type $<t, x, y>$
representing a sample of the movement of an object:

$<0, 10, 15>; <5, 15, 10>; <10, 25, 10>; <15, 30, 20>; <20, 15, 40>;$
$<25, 40, 25>; <30, 50, 30>; <35, 60, 35>; <40, 70, 30>; <45, 75, 15>$

(a) Assuming constant speed during time period $[t_{i-1}, t_i)$, draw the time—
speed chart corresponding to this sequence.
(b) Based on this chart, argue about possible noise that may have appeared
in the sequence.
(c) After removing noise, redraw the time—speed chart corresponding to
the revise sequence.
(d) Examine whether every single record of the revised sequence is
required in order to store the movement of the object or some records
are redundant (because they are inferred by others) and may be
removed without lowering the quality of information.

Ex. 3.2. Study the GPS dataset you recorded for the purposes of Chap. 1 (recall
Ex. 1.2 and Ex. 1.3) and try to (a) identify and remove noise and (b) identify and smoothen random errors following the ideas discussed in Sect. 3.2.
Ex. 3.3. On the cleansed dataset resulted by Ex. 3.2, try to identify trajectories following the ideas discussed in Sect. 3.3.
Ex. 3.4. Using the revised sequence of recorded positions of type $<t, x, y>$ of
Ex. 3.1, try to map-match the resulting trajectory on a road network consisting of five vertices, $v_1(5, 30), v_2(10, 10), v_3(60, 35), v_4(80, 0), v_5(55, 70)$,
and four edges, $e_1(v_1, v_2), e_2(v_2, v_3), e_3(v_3, v_4), e_4(v_3, v_5)$.

(a) Which are the road segments (edges) the object followed?
(b) Which positions were the most problematic and why?

For map-matching, you can choose among the techniques discussed in
Sect. 3.3.1, or even follow your intuition and imagination!

3.7   Bibliographical Notes

The reasons why time cannot be simply treated as yet another spatial dimension
have been extensively discussed in the spatiotemporal database literature, see e.g.
(Theodoridis et al. 1998; Güting et al. 2000; Koubarakis et al. 2003).
    One of the first models for moving objects was proposed in Wolfson et al. (1998,
1999). The so-called *Moving Objects Spatio-Temporal* (MOST) data model applies
for databases with *dynamic attributes*, i.e. attributes that change continuously as a
function of time, without being explicitly updated. This model enables the DBMS
to predict the future location of a moving object by providing a *motion vector*,

consisting of its location, speed and direction, for a recent period of time. As an extension to the abstract model of Güting et al. (2000), the concept of *spatio-temporal predicates* is introduced in Erwig and Schneider (2002). The goal was to investigate temporal changes of topological relationships induced by temporal changes of spatial objects. Further work on modeling includes (Su et al. 2001) where the authors focused on moving point objects and the inclusion of concepts of differential geometry (speed, acceleration) in a calculus-based query language. In Becker et al. (2004), a non-linear representation for moving objects was discussed in detail, while (Vazirgiannis and Wolfson 2001) considered movement in networks and some evaluation strategies.

Regarding the sampling and uncertainty of mobility data, (Pfoser and Jensen 1999) presented and formalized the problem in the spatiotemporal database domain. As far as it concerns data cleansing approaches, Yan et al. (2010) proposed a Gaussian kernel-based local regression model to smoothen GPS feeds. In an effort to handle the entire trajectory reconstruction process (i.e. both steps of data cleansing and trajectory identification), Marketos et al. (2008) presented a parametric online approach for distinguishing trajectories within a stream of incoming positions, while at the same time smoothing trajectories, avoiding redundancies, identifying and removing outliers (noise).

The network models for representing transportation networks (edge- vs. route-oriented) are discussed in Güting et al. (2006), Speicys and Jensen (2008). Segmenting trajectories into homogeneous pieces is discussed in Lee et al. (2007), Buchin et al. (2010), Panagiotakis et al. (2012). Regarding map-matching, Brakatsoulas et al. (2005) proposed a state-of-the-art topological technique, which can work either offline or online. Quddus et al. (2003) proposed a technique for replacing each position of the original trajectory by the point on the network that is the most likely position of the moving object.

Douglas-Peucker line simplification algorithm was proposed in Douglas and Peucker (1973). Top-Down Time Ratio (TD-TR) and Open Window Time Ratio (OPW-TR) algorithms for compressing spatiotemporal data were proposed in Meratnia and de By (2004). Potamias et al. (2006) proposed two algorithms, called *Thresholds* and *STTrace*, respectively, for online trajectory data compression. A trajectory simplification method aiming at Location-based social networking (LBSN) applications was proposed in Chen et al. (2009). A combined compression and map-matching approach was proposed in Kellaris et al. (2013).

Regarding the trajectory generators presented in Sect. 3.4, Theodoridis et al. (1999) proposed the *GSTD* data generator, which has been followed by a number of variations, e.g. (Pfoser and Theodoridis 2003), which supported obstacles infrastructure. *CENTRE* was proposed in Giannotti et al. (2005). On the other hand, for network-constrained movement, *Brinkhoff* was proposed in Brinkhoff (2002), *SUMO* was proposed in Krajzewicz et al. (2002), *GAMMA* was proposed in Hu and Lee (2005), *BerlinMOD* was proposed in Düntgen et al. (2008), *MWGen* was proposed in Xu and Güting (2012), and *Hermoupolis* was proposed in Pelekis et al. (2013).

References

Becker L, Blunck H, Hinrichs K, Vahrenhold J (2004) A framework for representing moving objects. In: Proceedings of DEXA

Brakatsoulas S, Pfoser D, Salas R, Wenk C (2005) On map-matching vehicle tracking data. In: Proceedings of VLDB

Brinkhoff T (2002) A framework for generating network-based moving objects. Geoinformatica 9(1):153–180

Buchin M, Driemel A, van Kreveld M, Sacristán V (2010) An algorithmic framework for segmenting trajectories based on spatio-temporal criteria. In: Proceedings of GIS

Chen Y, Jiang K, Zheng Y, Li C, Yu N (2009) Trajectory simplification method for location-based social networking services. In: Proceedings of LBSN

Douglas D, Peucker T (1973) Algorithms for the reduction of the number of points required to represent a digitized line or its caricature. Can Cartogr 10(2):112–122

Düntgen C, Behr T, Güting RH (2008) BerlinMOD: a benchmark for moving object databases. VLDB J 18(6):1335–1368

Erwig M, Schneider M (2002) Spatio-temporal predicates. IEEE Trans Knowl Data Eng 14(4):881–901

Giannotti F, Mazzoni A, Puntoni S, Renso C (2005) Synthetic generation of cellular network positioning data. In: Proceedings of GIS

Güting RH, Böhlen MH, Erwig M, Jensen CS, Lorentzos NA, Schneider M, Vazirgiannis M (2000) A foundation for representing and querying moving objects. ACM Trans Database Syst 25(1):1–42

Güting RH, de Almeida VT, Ding Z (2006) Modeling and querying moving objects in networks. VLDB J 15(2):165–190

Hu H, Lee DL (2005) GAMMA: a framework for moving object simulation. In: Proceedings of SSTD

Kellaris G, Pelekis N, Theodoridis Y (2013) Map-matched trajectory compression. J Syst Softw 86(6):1566–1579

Koubarakis M, Sellis TK, Frank AU et al (2003) Spatio-temporal databases—the CHOROCHRONOS approach. Springer, New York

Krajzewicz D, Hertkorn G, Rössel C, Wagner P (2002) SUMO (Simulation of Urban MObility): an open-source traffic simulation. In: Proceedings of MESM

Lee JG, Han J, Whang KY (2007) Trajectory clustering: a partition-and-group framework. In: Proceedings of SIGMOD

Marketos G, Frentzos E, Ntoutsi I, Pelekis N, Raffaetà A, Theodoridis Y (2008) Building real-world trajectory warehouses. In: Proceedings of MobiDE

Meratnia N, de By RA (2004) Spatiotemporal compression techniques for moving point objects. In: Proceedings of EDBT

Panagiotakis C, Pelekis N, Kopanakis I, Ramasso E, Theodoridis Y (2012) Segmentation and sampling of moving object trajectories based on representativeness. IEEE Trans Knowl Data Eng 24(7):1328–1343

Pelekis N, Ntrigkogias C, Tampakis P, Sideridis S, Theodoridis Y (2013) Hermoupolis: a trajectory generator for simulating generalized mobility patterns. In: Proceedings of ECML-PKDD

Pfoser D, Jensen CS (1999) Capturing the uncertainty of moving-object representations. In: Proceedings of SSD

Pfoser D, Theodoridis Y (2003) Generating semantics-based trajectories of moving objects. Comput Environ Urban Syst 27(3):243–263

Potamias M, Patroumpas K, Sellis TK (2006) Sampling trajectory streams with spatiotemporal criteria. In: Proceedings of SSDBM

Quddus MA, Ochieng WY, Zhao L, Noland R (2003) A general map matching algorithm for transport telematics applications. GPS Solutions 7(3):157–167

Speicys L, Jensen CS (2008) Enabling location-based services—multi-graph representation of transportation networks. Geoinformatica 12(2):219–253

Su J, Xu H, Ibarra O (2001) Moving objects: logical relationships and queries. In: Proceedings of SSTD

Theodoridis Y, Sellis TK, Papadopoulos AN, Manolopoulos Y (1998) Specifications for efficient indexing in spatio-temporal databases. In: Proceedings of SSDBM

Theodoridis Y, Silva JRO, Nascimento M (1999) On the generation of spatiotemporal datasets. In: Proceedings of SSD

Vazirgiannis M, Wolfson O (2001) A spatiotemporal model and language for moving objects on road networks. In: Proceedings of SSTD

Wolfson O, Sistla AP, Chamberlain S, Yesha Y (1999) Updating and querying databases that track mobile units. Distrib Parallel Dat 7(3):257–387

Wolfson O, Xu B, Chamberlain S, Jiang L (1998) Moving objects databases: issues and solutions. In: Proceedings of SSDBM

Xu J, Güting RH (2012) MWGen: a mini world generator. In: Proceedings of MDM

Yan Z, Parent C, Spaccapietra S, Chakraborty D (2010) A hybrid model and computing platform for spatio-semantic trajectories. In: Proceedings of ESWC

Chapter 4
Mobility Database Management

Adding temporal information, as an extra attribute in spatial databases, is not as straightforward as it may appear at a first glance. Time is not yet another dimension besides the two (or three, in some applications) spatial dimensions; monotonicity, for example, is a key difference. Could we adopt "as-is" methods and techniques for spatial databases, such as the ones outlined in Chap. 2? The answer is rather no, and this has been argued extensively in the spatiotemporal database literature. Therefore, novel data types (e.g. moving points), query processing techniques (e.g. "search for trajectories that 'entered' an area during a timeframe" or "search for trajectories that are 'similar' with respect to a reference trajectory") and indexing methods (most probably, extensions of the well-known R-tree) have been explored. This chapter surveys the above aspects, which are essential components of a database system targeting at efficiently handing mobility data. In particular, interesting location- and mobility-aware queries are overviewed. Then, at physical level, selected indexing and query processing techniques that have been designed to efficiently support the above models and query types are presented.

4.1 Location- and Mobile-Aware Querying

The nature of trajectory data provides us with the ability to query them with a variety of operators. Location-aware queries could be viewed with the perspective of instantiations of the dimensions below:

- *Time addressed*: past vs. present vs. future
- *Duration of query*: snapshot vs. continuous
- *Type of query*: range, nearest-neighbor (NN), distance, topological, directional, etc.
- *Type of reference object(s)*: stationary vs. moving
- *Type of data objects*: stationary vs. moving

N. Pelekis and Y. Theodoridis, *Mobility Data Management and Exploration*,
DOI 10.1007/978-1-4939-0392-4_4, © Springer Science+Business Media New York 2014

In other words, irrespective of the query type (it could be range, NN, etc.), the parameters taken into consideration to define a location-aware query include: (a) the time the query refers to (i.e. whether it refers to the past, the present or the (antici- pated) future), (b) the duration during which the query is valid (hence, we may define snapshot vs. continuous queries), and (c) the nature of the reference as well as the data objects, whether they are stationary or moving. The above discussion also infers that a MOD may consist of mobility data as well as (stationary) spatial data, e.g. points (POI) or regions of interest (ROI), and cross-over queries between the two datasets are also of interest.

Here are a few examples, according to the above taxonomy, based on the applica- tion scenario of vessels sailing on sea presented in Chap. 1:

– *Querying the past*:

i.   *Search for vessels that were located within 1 km distance from shore yester- day night*; query of type <Past, Snapshot, Distance, Stationary, Moving>.

ii. *Find the average speed and heading of vessels that were sailing south of island of Crete from 07:00 to 10:00 yesterday*; query of type <Past, Snapshot, Directional, Stationary, Moving>.

iii. *Search for vessels that were found close to each other, in a distance less than 1 n.m, yesterday, while in motion*; query of type <Past, Snapshot, Distance, Moving, Moving>.

– *Querying the present*:

iv. *How many vessels are located in port of Piraeus now?*; query of type <Present, Snapshot, Range, Stationary, Moving>.

v.   *What is the current vessel-to-vessel distance for all vessels sailing in the Euboea-Andros short passage now?*; query of type <Present, Snapshot, Distance, Moving, Moving>.

– *Querying the (anticipated) future*:

vi. *For the next 10 min of its trip, monitor the 3- nearest reefs that vessel XYZ is expected to approach during its movement*; query of type <Future, Continuous, NN, Moving, Stationary>.

vii. *Search for the biodiversity areas, vessel XYZ is expected to cross during the next hour of its trip*; query of type <Future, Snapshot, Range, Moving, Stationary>.

Moreover, the so-called *motion pattern queries* are time-ordered sequences of primitive queries, where time ordering could be either explicit or implicit. Continuing the previous series of example queries:

viii. *Search for ships that were located within 1 km distance from shore yesterday night, arrived at a port between 06:00 and 08:00 today, and stayed in the port for at least 1 hour*; this is a temporal sequence of three simpler queries.

In a different perspective, Table 4.1 classifies queries as location- vs. trajectory- oriented queries with the distinction emphasizing on the assumed model of

Table 4.1 Location- and trajectory-oriented queries

Query type		Examples
Location-oriented		Continuous nearest neighbor, sequenced route (trip planning), etc. queries
Trajectory-oriented	Coordinate-based	Range, NN, etc. queries
	Trajectory-based	Topological (enter, leave, cross, bypass, etc.), navigational (in front of, etc.), motion pattern, trajectory similarity, etc. queries
	Traffic analysis	Path, traffic, etc. queries

trajectory objects: the former consider movement as a sequence of sampled points whereas the latter consider movement as a continuous evolution (trajectory).

4.1.1 Location-Oriented Queries

This class of queries covers operations that are essentially spatial, though taking into consideration that the spatial objects are changing their locations with time. Typical examples include continuous nearest neighbor, sequenced route queries, etc.:

- *Continuous Nearest Neighbor (CNN) query*: a CNN query retrieves the nearest (among a set of candidate points) of every location on a polyline (actually, the trajectory of a moving object from a starting to an ending point). The result set is a set of pairs (*point*, *interval*), where *point* is the nearest point to the polyline during *interval*. Here is a typical example: "find the nearest gas stations during my route from A to B".
- *Sequenced Route (SR) query*: a SR query finds the shortest path from a starting s to an ending point e, visiting a sequence of facilities from a set of facility classes. The result set is a path from s to e satisfying the constraints set. Here is a typical example: "find the best route from A to B visiting an ATM, then a gas station, and then a restaurant".

Figure 4.1 exemplifies CNN and SR queries. Assume a pair (s, e) denoting the starting and ending point of a moving object as well as three sets of facilities, A_i, G_i, and R_i, for ATMs, gas stations, and restaurants, respectively. In a CNN query, we aim to find the nearest gas stations with respect to a (pre-defined) route from s to e (solid line in Fig. 4.1; for simplicity a line segment). In our example, the result is the set $\{(G_1, [t_s .. t_i]), (G_2, [t_i .. t_j]), (G_4, [t_j, t_e])\}$. On the contrary, in SR queries the result is the route from s to e itself, which is built to satisfy the constraints; in our example visiting an ATM, then a gas station, and then a restaurant. The result is the route $<s, A_1, G_2, R_2, e>$ (dashed line in Fig. 4.1). Note that although R_1 and A_3 are very close to the straight line from s to e, they are not included in the answer set due to the constraint enforced on the temporal order of the facilities.

Fig. 4.1 Examples of
location-oriented queries

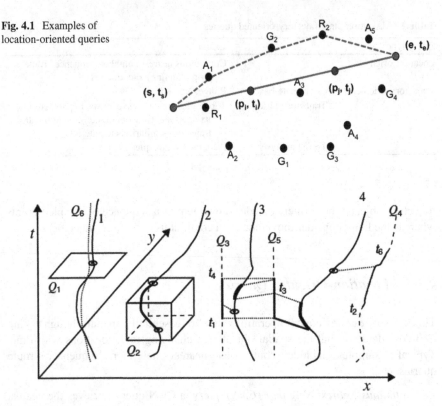

Fig. 4.2 Examples of trajectory-oriented queries

4.1.2 Trajectory-Oriented Queries

Focusing on this class, which is the most popular in MOD literature, we may consider *coordinate-based* queries (timeslice, range, nearest-neighbor, etc.), which are straightforward extensions of spatial queries in 3-dimensional space, as well as *trajectory-based* queries (topological, navigational, similarity, motion pattern, etc.), which are novel and emerging from the trajectory concept.

In coordinate-based queries, one is interested in filtering the trajectory database setting conditions on the (space- and/or time-) coordinates of the segments that compose the trajectories. For instance, in Fig. 4.2, we may consider:

- *Timeslice queries*: "find the locations of trajectories at a given timestamp" (denoted as Q_1 in Fig. 4.2).
- *Spatiotemporal range queries*: "find the portions of trajectories inside a given spatial region during a given time interval" (Q_2).
- *Nearest-neighbor queries* in three versions, *point NN*, *trajectory NN*, and *historical continuous point NN*: "find the trajectories that lie nearest to a given point

during a given time interval" (Q_3), "… nearest to a given trajectory during a given time interval" (Q_4), and "… nearest to a given point in every time instance of a given time interval" (Q_5), respectively.

On the other hand, in trajectory-based queries, it is essential to have knowledge of the entire trajectory (or, at least, a sub-trajectory of it) in order to be able to provide the answer. Again on the example of Fig. 4.2, we may consider *trajectory similarity queries* ("find the trajectories that are the most similar to a given trajectory", denoted as Q_6 in Fig. 4.2), as well as *topological* and *navigational queries* with respect to a stationary object ("find the trajectories that entered (crossed, left, bypassed, etc.) a given region during a given time interval" and "…were located west of (south of, etc.) a given region …", respectively) as well as their counterparts when the reference object is another trajectory ("find the trajectories that met (followed, were in front of, etc.) a given trajectory").

Moreover, combined coordinate- and trajectory-based queries are expected to be of great interest in MOD. Consider the following example: "find trajectories that entered a given region during a given time interval, stayed inside the region for a given time period, and then left the region at a speed higher than a given speed threshold". For sure, interesting query processing issues arise here.

Last but not least, an emerging family of trajectory-oriented queries is motivated by traffic analysis; examples include *path queries* ("find the trajectories that have moved along an entire path") and *traffic queries* ("find places on the road network where traffic jams appear"), etc.

As already mentioned, coordinate-based queries can be considered (at least in their definitions) as straightforward extensions of the respective queries on spatial databases (point, range, and NN queries) discussed in Chap. 2. On the other hand, trajectory-based queries are quite distinct from those studied in spatial databases. Hence, appropriate trajectory-based indexing and query processing techniques are required in order to support them (to be discussed in Sects. 4.2 and 4.3, respectively). Before this discussion, a note on uncertainty is necessary.

4.1.3   Querying Under Uncertainty

In our discussion so far, we considered that recorded information is trustworthy. However, this is not true: GPS devices introduce inherent error in measurements, as it was discussed in Chaps. 1 and 3. This fact introduces uncertainty in the query results. In the example illustrated in Fig. 4.3, assuming that the actual location of an object could be anywhere in a disk (uncertainty circle) around the recorded location, it is clear that the result set of e.g. a range query may include *false hits* (i.e. points recorded inside the query window but their actual location was outside) and may not include *missed hits* (i.e. points recorded outside the query window but their actual location was inside).

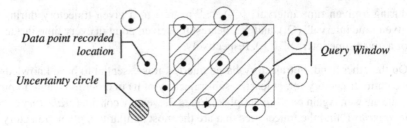

Fig. 4.3 The effect of uncertainty in spatial range queries

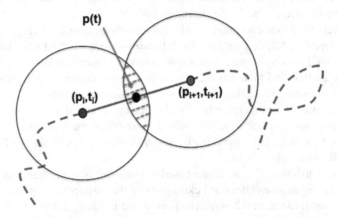

Fig. 4.4 The uncertainty area of a moving object at time t between two sampled points

Taking uncertainty into consideration, the members of a point dataset with respect to the window Q of a range query are distinguished in two classes, *MUST* and *MAY*, as follows:

– A point p is included in the *MUST* set with respect to Q if it sure that p is inside Q whatever was its actual location; in other words, the uncertainty circle of p is inside Q.
– A point p is included in the *MAY* set with respect to Q if it probable (but not sure) that p is inside Q; in other words, the uncertainty circle of p partially overlaps Q.

The implications of uncertainty in spatial locations appear also in trajectory databases. Let us refresh the discussion on trajectory reconstruction and the linear interpolation method provided in Chap. 3. Suppose we know moving object's maximal speed v and two recorded positions, $<p_i, t_i>$ and $<p_{i+1}, t_{i+1}>$. It is obvious that at any time $t \in (t_i, t_{i+1})$, the object is somewhere in the intersection of two disks, the first with center p_i and radius $r = v(t - t_i)$ and the second with center p_{i+1} and radius $r = v(t_{i+1} - t)$, as illustrated in Fig. 4.4. Note that the union of all lenses (see the dashed area) formed by the intersection of the two disks at all timestamps $t \in (t_i, t_{i+1})$ results in an ellipsis with p_i and p_{i+1} as its foci, which is the potential area of activity (PAA) of the moving object.

On the other hand, linear interpolation assumes that the object is located at a specific location $p(t)$, as discussed in Chap. 3. This conflict has an effect in querying mobility data when the inherent uncertainty needs to be taken into consideration. For instance, there could be uncertainty-aware variations of the queries presented earlier in this section, like for example, the topological query *inside* ("search for trajectories that were located inside a region during a time interval") could appear in two versions: *definitely-inside* and *possibly-inside*, where the difference is that in the former we are confident that the answers fulfill the topological constraint *inside* even if the actual movement was not the one produced by the linear interpolation but any movement resulting due to the uncertainty area, while in the latter we cannot have a warranty of this.

A method in order to support this type of uncertainty queries is to associate an uncertainty threshold r to the trajectory. In other words, each sampled position $<p, t>$ of the trajectory is transformed to a disk $<p \pm r, t>$, making the trajectory be modeled as a cylindrical volume in 3-dimensional space around the given trajectory. Continuing the previous example, if the cylindrical volume is entirely inside the given polygon, then the trajectory is in the *definitely-inside* answer set; on the other hand, if each disk composing the cylindrical volume at least overlaps the given polygon, then the trajectory is in the *possibly-inside* answer set.

4.2 Indexing Techniques for Mobility Data

R-tree is ubiquitous in spatial databases, as it was sketched in Chap. 2. A natural question that arises is whether it is portable to mobility data. In other words, at least for indexing purposes, could *time* be treated as yet another dimension additional to the two (or three, in some applications) spatial dimensions? In order to provide an answer to this question, let us put some specifications on the table:

(a) Both *space* and *time* should be treated as equally important dimensions.
(b) *Time* is not yet another dimension due to is special characteristics (consider e.g. its monotonicity).
(c) Density in 'spatiotemporal' space may significantly vary, from place to place and from time to time.

Linking this discussion with the classification of queries as coordinate- vs. trajectory-based, presented in Sect. 4.1, it appears that handling trajectories as 3- (or 4-) dimensional polylines may be expected to work for the former type of queries but it would for sure have drawbacks for the latter type. Indeed, a variety of spatiotemporal access methods for moving object trajectories have been proposed based on the R-tree. In the following paragraphs, we choose to present a few representatives for either unconstrained or network-constrained movement.

Fig. 4.5 Approximating a
trajectory by either a single
MBB or a sequence of MBBs
to be indexed in a 3D R-tree

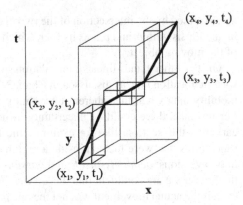

4.2.1   Indexing Trajectories in Free Space

3D R-tree. It is the earliest method for indexing trajectories of moving objects.
The 3D R-tree is a straightforward extension of the R-tree in 3-dimensional space,
considering $2+1$ (spatial and temporal, respectively) dimensions. As its name implies,
it treats time as an extra spatial dimension and it decomposes trajectories in
3-dimensional segments storing segments' Minimum Bounding Boxes (MBB) in its
nodes, as illustrated in Fig. 4.5. As it is obvious from Fig. 4.5, excessive dead space is
introduced to the structure since a polyline is approximated by a set of boxes, which,
nevertheless, would have been much larger should a single MBB be used instead.

Insertion in 3D R-trees is performed as in the original R-tree, which implies that
spatiotemporal proximity is of interest. Hence, the insertion process does not take
the monotonicity of time into special account. As expected, 3D R-tree turns out to
be efficient in answering coordinate-based (timeslice and range) queries, while if
fails to efficiently support trajectory-based queries. This is an expected behavior
since time dimension has no special treatment there.

Several techniques have tried to address the *trajectory preservation* and *skewed
data growth* problems that arise when indexing a MOD. The former refers to the
fact that traditional spatial indexing techniques, e.g. R-trees, prefer to group together
entities according to spatial proximity. So it is natural for portions of different trajectories instead of those of the same trajectory to be grouped together in the same
node of the index. However, this is a problem when trajectory-based queries are to
be processed. The second issue (skewed data growth) is inherent to the nature of
time dimension and it is due to its monotonicity. However, monotonous evolution of
information in one of the dimensions is far from the concept of spatial indexing
techniques. Concerning the above two issues, the TB-tree is one of the most successful methods for indexing trajectory data.

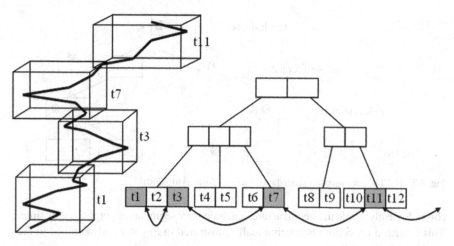

Fig. 4.6 The TB-tree indexing method

TB-tree. The TB-tree (Trajectory Bundle tree) maintains the 'trajectory' concept. Each node consists of segments of a single trajectory and nodes corresponding to the same trajectory are linked together in a chain. As such, the entire trajectory of a moving object or even sub-trajectories of it can be reconstructed by visiting the nodes of the chain; hence, it turns out to be effective for coordinate-based as well as trajectory-based queries.

In detail, TB-tree is a height-balanced tree with the index records in its leaf nodes; leaf nodes are of the form (*MBB, Orientation*), where *MBB* is the 3-dimensional bounding box of the 3-dimensional line segment belonging to an object's trajectory (recall that a leaf node contains entries of a single trajectory only) and *Orientation* is a flag used to reconstruct the actual 3-dimensional line segment inside the MBB among four different alternatives that could exist. Since, by definition, each leaf node contains entries of a single trajectory, the object identifier (*id*) needs to be stored only once, in the leaf node header. Like R-tree, internal and leaf node MBBs belonging to the same tree level are allowed to overlap. Each internal or leaf node in the tree corresponds to a physical disk page (or disk block, which is the fundamental element on which the actual disk storage is organized) and contains between *m* and *M* entries (*M* is the node capacity—fanout—and *m* in the case of TB-tree is set to 1). For each trajectory, a double linked list connects leaf nodes together to reconstruct the entire trajectory, as illustrated in Fig. 4.6.

Regarding the maintenance of the structure and unlike R-tree, the TB-tree insertion algorithm is not based upon the spatial and temporal relations of moving objects but it rather relies only on object's *id*. As soon as new line segments are inserted, the algorithm searches for the leaf node containing the last entry of the same trajectory, and simply inserts the new entry into that node, thus forming leaf nodes that exclusively contain line segments from a single trajectory. If the leaf node is full, then a new one is created and is inserted in the right-end of the tree; due to the monotonicity of time, this strategy ensures that trajectories are organized

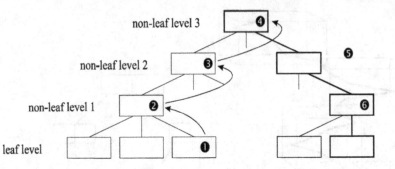

Fig. 4.7 The TB-tree insertion procedure (*source*: Pfoser et al 2000)

chronologically inside the tree structure, e.g. trajectory segments are organized by time. This insertion procedure is schematically illustrated in Fig. 4.7 and sketched in the pseudocode that follows.

Algorithm TB-tree-Insert

Input: TB-tree node N, entry e

Output: (none)

```
1.   N' = TB-tree-FindNode (N, e)
2.   IF N' was found
3.      IF N' has space
4.         insert e into N'
5.      ELSE // N' is full of entries
6.         create new leaf node New
7.         insert e into New
8.   ELSE // N' was not found
9.      create new leaf node New
10.     insert e into New
```

Algorithm TB-tree-FindNode

Input: TB-tree node N, entry e

Output: TB-tree node N'

```
1.   IF N is a non-leaf node
2.      FOR each entry e' in N
3.         IF e'.MBB overlaps e.MBB
4.            call TB-tree-FindNode (e'.ptr, e)
5.   ELSE // N is a leaf node
6.      FOR each entry e' in N
7.         IF e'.MBB is connected to e.MBB
8.            N' = N
9.            return N'
```

Furthermore, in order to support high insertion rates, the TB-tree may use an in-memory hashed front-line structure, which maintains tuples of the form $\langle id, p_{curr}, ptr_{curr} \rangle$ consisting of object's id, its most recently recorded position $p_{curr} = \langle t_{curr}, x_{curr}, y_{curr} \rangle$, and a pointer ptr_{curr} pointing to the leaf node containing p_{curr}.

4.2.2 Indexing Network-Constrained Trajectories

In case of network-constrained movement (as in transportation networks), one of the earliest approaches for indexing trajectories exploited on dimensionality reduction. With this approach, 3-dimensional trajectories are translated into two dimensions $(p\text{-}, t\text{-})$, where $p\text{-}$ dimension is a mapping of 2-dimensional network to a single dimension using a space filling curve, such as Hilbert curve. Of course, in such a scenario, mappings for the queries should be devised as well.

In the actual 3-dimensional space, indexing techniques for network-constrained trajectories include variations of the ubiquitous R-tree in combination with appropriate handling of the underlying network. FNR-tree and MON-tree are among the most cited ones.

FNR-tree and **MON-tree**. Both the FNR-tree (Fixed Network-constrained R-tree) and the MON-tree (Moving Objects in Networks tree) have been designed to support network-constrained movement in two layers: one the one hand, a spatial (2-dimensional) R-tree index organizes the underlying network and, on the other hand, a forest of (1-dimensional, in the case of FNR-tree; 2-dimensional, in the case of MON-tree) R-tree indexes organize the movement of objects on the network.

In particular, as illustrated in Fig. 4.8, the transportation network and the movement along it are reflected into an architecture, consisting of one 2-dimensional R-tree along with a forest of several 1-dimensional R-trees, respectively.

In particular, the spatial (2-dimensional) R-tree organizes the spatial information about the line segments of the network. Each leaf node of the 2-dimensional R-tree contains a pointer to the root of a 1-dimensional R-tree; the latter organizes the time intervals (note that no spatial information is included here) that a moving object was moving on the network.

Regarding the maintenance of the structure, as soon as a new position arrives, the insertion algorithm propagates down the 2D R-tree searching for the line segment s of the network on which the movement has been performed. Then, it follows the pointer to the root of the respective 1D R-tree and inserts into that structure the time interval of the movement along s. Note that since the transportation network does not change, or at least changes very rarely, we can safely assume that the 2D R-tree remains static during the lifetime of the FNR-tree.

As derived from the above discussion (and illustrated in more detail in Fig. 4.8b), each 1-dimensional R-tree of the FNR-tree is in charge of the movement performed on the (sub-)network contained in the MBR of the corresponding 2-dimensional R-tree leaf node. However, this modeling is the main disadvantage of the FNR-tree; since

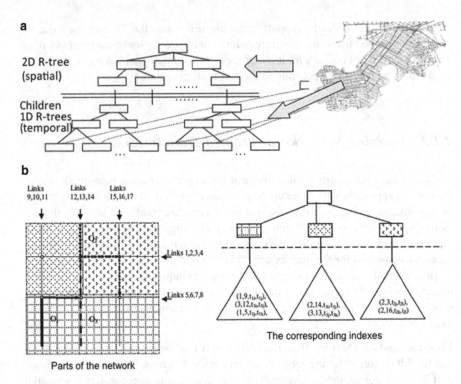

Fig. 4.8 The FNR-tree indexing method: (**a**) the general concept; (**b**) a detail of the structure

only the time intervals of the movement are stored in the 1-dimensional R-tree, it is assumed that the objects move along the entire length of a network edge. This limitation is addressed by the MON-tree. Instead of maintaining a forest of 1-dimensional R-trees for the movement, a forest of 2-dimensional R-trees is used providing much more flexibility in movement modeling.

PARINET. Unlike previous techniques, PARINET is a grid partitioning approach maintaining B+-tree indexes on time intervals for the (sub-)network covered by each partitioning. Actually, for the purposes of PARINET, every trajectory is decomposed in a set of $\{t_{id}, r_{id}, [pos_1, pos_2], [t_1, t_2]\}$ denoting that a trajectory identified by t_{id} moved along a road identified by r_{id}, from pos_1 to pos_2 of the road, during time interval $[t_1, t_2]$. With the above modeling into consideration, the index is built in three phases:

– Trajectory portions are partitioned according to the road identifiers they move along.
– Partitions are sorted on the time intervals.
– Each partition is indexed using a B+-tree on the time intervals contained in the partition.

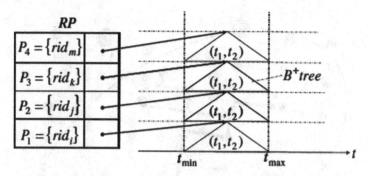

Fig. 4.9 The PARINET index structure (*source*: Sandu Popa et al. 2011)

Figure 4.9 illustrates an example of PARINET index structure. A table on the left (RP for Road Partitioning) contains one entry for each cluster keeping some basic information on the partitioning: a list of road identifiers contained in the partition and a pointer to the respective B^+-tree. The set of B^+-trees (on the right) is built upon the time intervals contained in the cluster.

PARINET is capable of answering range and NN queries over trajectories as well as dedicated *path queries*. For instance, PARINET efficiently retrieves all objects that have traversed all road segments composing a query path Q.

4.3   Query Processing Techniques

With the knowledge of indexing techniques we have in our hands, R-trees and family for spatial (2-dimensional data), 3D R-trees, TB-trees, etc. for spatiotemporal (3-dimensional data), we revisit the interesting queries discussed in Sect. 4.1 and present methods for their efficient processing.

4.3.1   Processing Location-Oriented Queries

CNN queries. In the same fashion with NN queries discussed in Chap. 2, CNN queries are efficiently processed using (2-dimensional) R-trees to prune the search space (i.e. the set of candidate facilities, e.g. gas stations). Specifically, starting from the root of the R-tree, the index is traversed according to the principle that we visit R-tree nodes only when they may contain qualifying data. Of course, whether a node may contain qualifying data or not is a heuristic decision, so there is room for variations of CNN search algorithms with respect to the specific heuristic followed.

SR queries. In this query, the result is the shortest path from a starting s to an ending point e, visiting a sequence of facilities from a set of facility classes. actually, it is a variation of the *travelling salesman problem* (TSP), thus it shares with TSP a

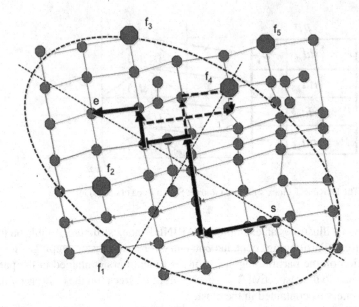

Fig. 4.10 Constrained routing on a road network

high complexity. Nevertheless, heuristic solutions can be found in polynomial time. For processing SR queries, one should distinguish between two cases: free (unconstrained) vs. network-constrained movement. In the former case, the Euclidean distance is assumed between all points in the space, therefore, pruning of candidate results may be efficiently performed if they are organized in a spatial index, such as the R-tree. On the other hand, network-constrained movement assumes distances corresponding to path lengths on the network, which are not supported by R-trees. The Constrained Routing (CR) algorithm addresses this issue by using the R-tree to limit the candidate set (note that CR solves a special case of SR query, where there is only one facility class in the constraint).

Figure 4.10 illustrates an example of performing CR on a road network. Assuming we route from s to e, the solid black line is the shortest path (taking no constraints into consideration). Using this path as query object, CR searches for the 1-NN (in Euclidean terms) among the set of facilities f_i (green/dark octagons) that are organized in a spatial R-tree index. The result (f_4) is used as a first candidate result for CR, hence we route from s to f_4 and, then, from f_4 to e. The length d of the total path is an upper bound for the final result, so we can use this distance to prune the facilities that are outside the elliptical area defined by the pair of foci (s, e) and distance d performing a range search over the set of facilities. (In fact, since facilities are organized in an R-tree, the ellipse is approximated by its MBR and a typical rectangular range search is performed.) The result set ($\{f_2, f_3, f_4\}$ in our example) is the set of candidates for CR. For each of them, the appropriate shortest paths are calculated in order to find the one that minimizes the overall distance (perhaps, f_2 in our example).

The pseudocode of the CR algorithm appears below:

Algorithm ConstrainedRouting

Input: road network G, starting point s, ending point e, R-tree root R

Output: selected facility f, selected path Pf

1.   Pf = ShortestPath (G, s, e)
2.   f = R-tree-based-NN-search (R, P, 1)
3.   Pf = ShortestPath (G, s, f) & ShortestPath (G, f, e)
4.   Fcandidates = R-tree-based-Range-search (R, MBR(P))
5.   FOR each entry f' in Fcandidates
6.   Pf' = ShortestPath (G, s, f') & ShortestPath (G, f', e)
7.   IF length(Pf') < length(Pf)
8.   f = f'; Pf = Pf'
9.   return (f, Pf)

4.3.2 Processing Trajectory-Oriented Queries

Aiming at the past vs. the present and (anticipated) future makes a big difference during query processing. When querying the past, the whole information is in our hands and scalability is actually the most critical issue—how to efficiently support queries in large volumes of data. On the other hand, when querying the present, the main issues are that data is continuously changing, real-time query support is required, and index structures should be update-tolerant. Furthermore, when querying the future, the main issue is that the movement should be somehow predicted and, obviously, this prediction remains valid only for a limited time horizon. Focusing on past, processing coordinate-based queries is a straightforward instantiation in the 3-dimensional space of the respective R-tree-based algorithms defined in multi-dimensional space. Assuming the trajectories are indexed in a 3D R-tree (see Fig. 4.5 and recall the discussion about R-trees in Chap. 2), typical examples include the range (timeslice, included) and NN queries below.

Range queries. The range search algorithm is based on the standard R-tree search algorithm propagating down the tree by filtering out nodes that could not contain entries that overlap the query window Q. In particular, the algorithm starts by visiting the tree root, checking which MBRs of the root entries overlap Q. For those entries, the algorithm follows the pointer to the corresponding child node, where it repeats recursively the same task. If a leaf node is reached, its entries are compared with Q and, if they overlap with Q, the algorithm reports their identifiers.

NN queries. At least two variations of k-NN search can be defined: the reference being a stationary object (usually, a point) or a trajectory. Extending the respective NN search algorithms proposed for R-trees in the spatial database domain, depth-first and best-first variations have been proposed. The best-first NN search algorithm works as follows: aided by a priority queue, in which the (node or leaf) entries of the tree nodes are stored in increasing order of their distance from the query object, at each tree node, the algorithm checks whether the lifetime of an entry overlaps the time period of the query, calculating at the same time its distance from the query object (so as the entry to be stored appropriately in the priority queue). At each algorithm's iteration, the first entry is requested from the queue until a leaf entry is found, which is then reported as the query result. The algorithm is incremental in the sense that the k-th NN can be obtained with very little additional work once the $(k-1)$-th NN has been found. k-NN search algorithms for moving objects in road networks have been also proposed. In this case, neighborhood is computed according to the network rather than the Euclidean distance.

Similarity queries. A modification of the NN query presented above is the so-called similarity search, formulated as follows: *given a trajectory dataset D, a reference trajectory T, a dissimilarity function dis() between two trajectories, and a distance threshold θ, find all trajectories $T_i \in D$ where dist(T_i, T)$\leq \theta$*. Since trajectories can be viewed as (2-dimensional) time series, several proposals adapt solutions from the time series domain while others face the problem as an extension of the NN search and provide solutions based on the NN query processing in trajectory databases discussed in the previous paragraph. A variation of similarity search as defined above, is the so-called *k-MST search* (i.e. k- most similar trajectory search) where among all trajectories $T_i \in D$, it is only the top-k with respect to their (dis) similarity from T that are reported.

(Distance) join queries. In accordance with the previous queries, join queries between two trajectory datasets (or in a single dataset, in case of self-join) extend join algorithms proposed for R-trees indexing spatial data (points or regions). Actually, the problem is formulated as follows: *given two trajectory datasets D_1 and D_2 and a distance threshold θ, find all pairs (l_1 l_2), where l_1 (l_2) is sub-trajectory of trajectory T_1 (T_2, respectively), $T_1 \in D_1$, $T_2 \in D_2$, such that their maximum (Euclidean) distance is less than θ*. An efficient solution to the special case of the problem where (l_1, l_2) is a pair of line segments, proposes treating trajectory segments as simple line segments in 3- dimensional space, indexing the two datasets in 3-dimensional R-trees, and employing R-tree spatial join algorithm to perform synchronized traversal of the two R-trees.

Assuming an index different from the 3D R-tree, query processing relies on the peculiarities of the specific index. For example, processing combined (coordinate- and trajectory-based) queries in the TB-tree works as follows: using the coordinate-based constraints of the query we perform a typical R-tree-style range search in the TB-tree. When we reach the leaf level, we will have retrieved a few segments from the trajectories that belong to the answer set. In a second step, we find the evolution of those trajectories, i.e. the segments that connect to the segments found in order to satisfy the second part of the query (the trajectory-based constraints). To do this, we follow the double linked lists that link the leaf nodes of the structure (see Fig. 4.6). The pseudocode for combined search in TB-trees appears below:

Algorithm TB-tree-based-CombinedSearch

Input: TB-tree node N, query window Q, query constraint C

Output: answer set S

```
1.   IF N is a non-leaf node
2.      FOR each entry e in N
3.         IF e.MBB overlaps Q
4.            CALL TB-tree-based-CombinedSearch (e.ptr, Q, C)
5.   ELSE // N is a leaf node
6.      FOR each entry e in N
7.         IF e.MBR overlaps Q AND e.T_id not in T_id
8.            Add e.T_id to T_id
9.            CALL DetermineTrajectory (N, e, C)
```

Algorithm DetermineTrajectory

Input: TB-tree node N, TB-tree entry e, query constraint C

```
1.   IF e' satisfies C
2.      Add e' to S
3.   FOR each entry e' following e in the forward linked list starting from done N
4.      IF e' satisfies C
5.         Add e' to S
6.   FOR each entry e' preceding e in the backward linked list starting from node N
7.      IF e' satisfies C
8.         Add e' to S
```

4.4   Benchmarks

Managing mobility data, irrespective of whether the focus is on historical movement (trajectories) or current and anticipated future locations, has been addressed by dozens of methods and techniques in the recent years. However, in order for these techniques to find their place in the market, they should 'convince' that they are efficient and flexible enough under varying settings and circumstances that might appear e.g. in LBS/LBSN applications. For instance, requests like "where are my friends", "navigate me to my closest friend", and "among my friends, who was closest to me yesterday" should be processed efficiently under various workloads. This is exactly the objective of benchmarks: to define a set of representative operations (queries and updates) that should be addressed efficiently.

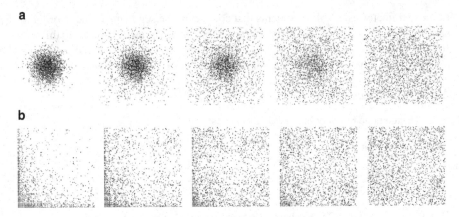

Fig. 4.11 The effect of data distribution in the evolution of moving objects: (**a**) movements starting with Gaussian distribution; (**b**) movements starting with skewed distribution

Benchmark parameters usually include:

- *Dataset configuration*: dataset type (real vs. synthetic), size, and distribution.
- *Index workload*: index size, maintenance cost for bulk loading, inserting, deleting, and updating entries.
- *Query workload*: cost (usually, I/O) for performing typical queries.

The first parameter (dataset configuration) involves real datasets for the purposes of proof-of-concept and synthetic datasets generated by appropriate tools (discussed in Chap. 3), mainly for the purposes of scalability and stress testing. It is obvious that a variety of space distributions should be covered since the performance of e.g. an access method may be significantly affected by the data distribution, especially in the case of mobility data where not only the original distribution but also its evolution is of interest. For example, as illustrated in Fig. 4.11, a population of moving objects may start evolving from different starting distributions (Gaussian and skewed, respectively) and, due to entropy of movement, in both cases result in an almost uniform distribution as time passes. (The screenshots of Fig. 4.10 were taken by GSTD generator, which was presented, among other spatiotemporal data generators, in Chap. 3.)

DynaMark and BerlinMOD are the benchmarks most related to the context of mobility data. DynaMark focuses on database servers supporting Location-Based Services (LBS). Simulating the movement of an arbitrary number of mobile users, it evaluates update as well as search costs. Update cost includes the cost of inserting and updating a user's location; search cost includes the cost of processing spatial queries that are core in LBS, such as proximity, k-NN, and sorted distance query. Cost measure is CPU time, as a main-memory index is assumed.

On the other hand, BerlinMOD is more pragmatic since it is based on real MOD system, called SECONDO (to be overviewed in Chap. 5). BerlinMOD simulates trajectories of vehicles moving on Berlin's road network; the simulated data contains spatial, temporal as well as thematic properties. The scenario is the following: there is a road network modeled as a set of nodes (2-dimensional points, representing street crossings and dead ends) and a set of edges (2-dimensional lines) connecting nodes, with a cost value assigned to each (representing free flow travel time). On this network, several types of persons' trips are modeled, including home—work—home trips in weekdays and other activity trips in evenings and weekends. Thus, 'home', 'work', 'neighborhood', etc. are key terms in the simulation of movements. In this setting, an extensive set of various types of queries (from non-spatial and simple spatial to sophisticated range, NN and other trajectory-oriented queries) is proposed, which evaluate the functionality of a MOD engine. To give the gist of BerlinMOD, below are (rephrased) examples of spatiotemporal queries included in the benchmark:

- *What are the pairs of vehicles that have ever been as close as 10 m or less to each other?*
- *When and where did the vehicles from a given set S meet other vehicles (distance <3 m)?*
- *Which vehicles passed a point from a given set P at a time instant from a given set TI?*
- *Which vehicles met at a point from a given set P at a time instant from a given set TI?*
- *Which vehicles travelled within a region from a given set R during the time periods from a given set TP?*
- *List the pairs of vehicles, the first from a given set S1, the second from a given set S2, where the corresponding vehicles are both present within a region from a given set R during a time period from a given set TP, but do not meet each other there and then.*
- *For each vehicle from a given set S and each time period from a given set TP: which are the 10 vehicles that are closest to that vehicle at all times during that period?*
- *Create ten disjoint groups of ten vehicles each. For each pair of such groups and each time period from a given set TP: report the vehicle(s) within the given first group of ten vehicles, having the minimum aggregated distance from the given other group of ten vehicles during the given period.*

Apart from query response times on the set of queries defined in the benchmark, BerlinMOD also counts statistics for the time required to build the database and the indexes, the disk size of the database, etc. Thus, it offers tools for benchmarking a MOD system by itself or even comparing it with SECONDO.

4.5 Summary

Due to the explosion of mobile devices, positioning technologies and the low data storage cost, one of the most important assets of knowledge intensive organizations working with movement data (from transportation engineering studying vehicle position data to social sciences and zoology studying human and animal movement data, respectively) is the mobility data itself. Having spatial database research results as starting point, research in mobility data management has resulted in a thesaurus of algorithms, methods and techniques, from indexing methods and query processing algorithms to handling uncertainty information that is inherent in mobility data sue to their collection mechanism.

These results are essential components, building blocks for providing mobility data management functionality in real-world systems, either prototype efforts or extensions of existing DBMS. This is exactly the topic to be discussed in the chapter that follows.

4.6 Exercises

Ex. 4.1. Develop a toy spatial database consisting of vessel routes (sequences of $<p_i, t_i>$ tuples), land (2-dimensional polygons), ports (2-dimensional rectangles), reefs (2-dimensional points), and biodiversity areas (2-dimensional polygons). Try to implement in SQL as many queries as possible from the 8 query examples listed in Sect. 4.1.

Ex. 4.2. According to the taxonomy of location-aware queries presented in Sect. 4.2, a query is defined as a vector in 5-dimensional space: <*time addressed, duration of query, type of query, type of reference object(s), type of data objects*>. Do you expect that every possible combination of instances in the above dimensions is feasible? Can you define counterexamples, i.e. instantiations of the 5 dimensions that cannot appear conjunctively?

Ex. 4.3. On the GPS dataset you recorded for the purposes of Chap. 1 (recall Ex. 1.2 and Ex. 1.3), simulate an example of (a) timeslice, (b) range, (c) NN, (d) topological, and (e) directional query.

Ex. 4.4. Explain the effect of uncertainty discussed in Sect. 4.1.3 on the queries you studied in Ex. 4.1 or Ex. 4.3.

Ex. 4.5. Refresh the discussion on TB-tree provided in Sect. 4.2.1. What exactly is the purpose of *Orientation* flag maintained at leaf nodes along with an MBB? Which are the four possible alternatives?

Ex. 4.6. The TB-tree insertion algorithm relies on time monotonicity, in other words, it assumes that every new position that arrives has a timestamp more recent than all timestamps already stored in the structure. What would happen if this were not the case? (Let, for instance, an application scenario where messages are delayed due to malfunction in the network communication.) Revise the algorithm to be insensitive to this assumption.

Ex. 4.7. Refreshing the discussion on FNR-tree in Sect. 4.2.2, does its insertion algorithm have the same drawback with TB-tree regarding the time monotonicity assumption? If no, why does it happen? If yes, how would you fix it?

Ex. 4.8. The range search algorithm sketched in Sect. 4.3.2 assumes 3D R-tree index for trajectories. Provide an alternative algorithm based on TB-tree. Can you argue in favor of the statement that TB-tree is not as appropriate as the 3D R-tree for range queries?

Ex. 4.9. Develop a toy spatial database consisting of vehicle routes (sequences of $<p_i, t_i>$ tuples), POI (2-dimensional points), and ROI (2-dimensional rectangles). Try to implement in SQL as many queries as possible from the list of BerlinMOD queries that appear in Sect. 4.4.

4.7 Bibliographical Notes

The taxonomy of location-aware queries based on five dimensions <*time addressed, duration of query, type of query, type of reference object(s), type of data objects*> was proposed in Mokbel and Aref (2007). Similar classifications of queries to that of Table 4.1 appear in Pfoser et al. (2000), Pfoser (2002). The distinction between coordinate- and trajectory-based queries is proposed in Pfoser et al. (2000). Motion patterns queries were proposed in Hadjieleftheriou et al. (2005) and further studied in Vieira and Tsotras (2011). Pelekis et al. (2007) discussed several variations of similarity queries: spatiotemporal, spatial only, with respect to derived information, such as speed and direction, etc.

The effect of uncertainty in spatial (range, etc.) queries is studied in Yu and Mehrotra (2003), from which is the classification of answers in *MUST* and *MAY* sets presented in Sect. 4.1.3, also in Frentzos et al. (2009), Frentzos et al. (2013). Trajcevski et al. (2004), de Almeida and Güting (2005b), and Tao et al. (2005) propose methods to handle uncertainty in MODs.

The ubiquity of R-trees in spatial databases has been documented in Manolopoulos et al. (2005). The 3D R-tree for the purposes of spatiotemporal indexing was proposed in Theodoridis et al. (1996) while it was adapted to organize trajectories of moving objects in Pfoser et al. (2000). Specifications for efficient indexing in spatiotemporal databases are discussed in Theodoridis et al. (1998). The overhead introduced by representing trajectory segments as MBBs in a R-tree like structure was studied in Hadjieleftheriou et al. (2002). Translating 3-dimensional trajectories into two dimensions using Hilbert curve and then indexing using (2-dimensional) R-trees was proposed in Pfoser and Jensen (2003).

As for the techniques discussed in Sect. 4.3, the TB-tree, the FNR-tree, the MON-tree, and PARINET were proposed in Pfoser et al. (2000), Frentzos (2003), de Almeida and Güting (2005a), and Sandu Popa et al. (2011), respectively. MV3R-tree (Tao and Papadias 2001) is another efficient proposal for indexing the past movement of mobility data, consisting of a multi-version R-tree (extending the idea of multi-version B-tree) and a small auxiliary 3D R-tree pointing to the leaf nodes of the former. The intuition behind the method is that timestamp queries can be addressed to the multi-version R-tree, whereas interval queries are supported by the auxiliary 3D R-tree. A recent approach in trajectory indexing includes TrajStore

(Cudre-Mauroux et al. 2010), which is actually a storage scheme consisting of distinct spatial and temporal indexes as well as a clustering mechanism that directs spatially co-located sub-trajectories into cells (collections of pages on disk).

Also in the case of indexing current locations of moving objects, indexing techniques are usually variations of R-trees or B$^+$-trees. Examples of the first class include the TPR-tree (Šaltenis et al. 2000), its descendant, TPR*-tree (Tao et al. 2003a), and the R^{PPF}-tree (Pelanis et al. 2006), which is a hybrid solution for both past and present movement. In particular, the Time Parameterized R-tree (TPR-tree) builds MBRs upon current locations using linear functions of time, which are organized in a R-tree (actually, R*-tree). TPR-tree works similarly to R-tree in update and search operations with the main difference that the penalty metric of R-tree (node MBR enlargement) is generalized to being an integral over a time period ranging from present time until a number of time units in the future. On the other hand, B$^+$-tree variations include the B^x-tree (Jensen et al. 2004) and the ST^{2}B-tree (Chen et al. 2008). Specifically, the B^x-tree partitions time into intervals of a predefined duration, which is the maximum duration in-between two updates of any object. For each interval, a number of phases (and corresponding index partitions) is defined. An object is inserted into an index partition according to the object's insertion time. As time passes, partitions expire, and new partitions are created receiving the orphan objects of expiring partitions.

CNN search was first introduced by Sistla et al. (1997). Efficient CNN query processing algorithms using R-trees were proposed in Song and Roussopoulos (2001), Tao et al. (2002a). Variations of CNN queries include CNN monitoring in the Euclidean space (Mouratidis et al. 2005) or network-constrained space (Mouratidis et al. 2006), and continuous visible NN queries (Gao et al. 2009). Sequenced route queries in the Euclidean space and in a road network were studied in Sharifzadeh et al. (2008) and Chen et al. (2011), respectively. Efficient processing of a variation of SR, called trip planning, was proposed in Li et al. (2005). The CR algorithm sketched in Sect. 4.3.1 was proposed in Frentzos et al. (2007c) under the name "In-route-find-the-nearest" and is based in the Incremental Euclidean Restriction (IER) algorithm proposed by Papadias et al. (2003) for efficient NN query processing in network-constrained environment.

The trajectory-oriented NN search algorithms sketched in Sect. 4.3.2 for movement in free space and under network constrains were proposed in Frentzos et al. (2007b) and Shahabi et al. (2003), respectively. Frentzos et al. (2007a) and Tiakas et al. (2009) studied trajectory similarity search in free space and under network constrains, respectively. Trajectory similarity search solutions based on time series are proposed in Vlachos et al. (2002), Chen et al. (2005). Distance join query processing in trajectory databases was studied in Bakalov et al. (2005), Arumugam and Jermaine (2006). Cost and selectivity estimation models in spatiotemporal (not necessarily, trajectory) databases are studied in Choi and Chung (2002), Tao et al. (2002b), Tao et al. (2003b).

Early spatial database benchmarks include Sequoia 2000 (Stonebraker et al. 1993) and Paradise (Patel et al. 1997), both based on NASA satellite data. However, these benchmarks are focused on the characteristics of a spatial database when used in the environment required for earth scientists. During the last decade, the research interest has focused on spatiotemporal and moving object databases. In this domain, the earliest approach is found at Theodoridis (2003) where a set of database queries appropriate for the purposes of location-based services (LBS) was proposed. COST benchmark (Jensen et al. 2006) evaluated three spatiotemporal indexes, namely the TPR-, TPR*-, and B^x-tree. DynaMark and BerlinMOD, mentioned in Sect. 4.4, were proposed in Myllymaki and Kaufman (2003) and Düntgen et al. (2009). Very recently, a new benchmark appeared, called Jackpine (Ray et al. 2011), which includes both micro benchmarks and macro workload scenarios. The micro benchmark component tests basic spatial operations in isolation; it consists of queries based on the 9-intersection model of topological relations proposed in Egenhofer and Herring (1991) as well as queries based on spatial analysis functions. Each macro workload includes a series of queries that are based on a common spatial data application. These macro scenarios include map search and browsing, geocoding, reverse geocoding, flood risk analysis, land information management and toxic spill analysis.

References

de Almeida VT, Güting RH (2005a) Indexing the trajectories of moving objects in networks. GeoInformatica 9(1):33–60

de Almeida VT, Güting RH (2005b) Supporting uncertainty in moving objects in network databases. In: Proceedings of ACM-GIS

Arumugam S, Jermaine C (2006) Closest-point-of-approach join for moving object histories. In: Proceedings of ICDE

Bakalov P, Hadjieleftheriou M, Keogh E, Tsotras V (2005) Efficient trajectory joins using symbolic representations. In: Proceedings of MDM

Chen L, Ozsu MT, Oria V (2005) Robust and fast similarity search for moving object trajectories. In: Proceedings of SIGMOD

Chen, S, Ooi BC, Tan KL, Nascimento MA (2008) ST^2B-tree: a self-tunable spatio-temporal B+−tree index for moving objects. In: Proceedings of SIGMOD

Chen H, Ku WS, Sun MT, Zimmermann R (2011) The partial sequenced route query with traveling rules in road networks. GeoInformatica 15(3):541–569

Choi Y, Chung C (2002) Selectivity estimation for spatio-temporal queries to moving objects. In: Proceedings of SIGMOD

Cudre-Mauroux P, Wu E, Madden S (2010) TrajStore: an adaptive storage system for very large trajectory data sets. In: Proceedings of ICDE

Düntgen C, Behr T, Güting RH (2009) BerlinMOD: a benchmark for moving object databases. The VLDB Journal 18(6):1335–1368

Egenhofer MJ, Herring J (1991) Categorizing binary topological relations between regions, lines and points in geographic databases. Technical Report, University of Maine

Frentzos, E. (2003) Indexing objects moving on fixed networks. In: Proceedings of SSTD

Frentzos E, Gratsias K, Theodoridis Y (2007a) Index-based most similar trajectory search. In: Proceedings of ICDE

Frentzos E, Gratsias K, Pelekis N, Theodoridis Y (2007b) Algorithms for nearest neighbor search on moving object trajectories. Geoinformatica 11(2):159–193

Frentzos E, Gratsias K, Theodoridis Y (2007c) Towards the next-generation of location based services. In: Proceedings of W2GIS

Frentzos E, Gratsias K, Theodoridis Y (2009) On the effect of location uncertainty in spatial querying. IEEE Transactions on Knowledge and Data Engineering 21(3):366–383

Frentzos E, Pelekis N, Giatrakos N, Theodoridis Y (2013) Cost models for nearest neighbor query processing over existentially uncertain spatial data. In: Proceedings of SSTD

Gao Y, Zheng B, Lee WC, Chen G (2009) Continuous visible nearest neighbor queries. In: Proceedings of EDBT

Hadjieleftheriou M, Kollios G, Tsotras VJ, Gunopulos D (2002) Efficient indexing of spatiotemporal objects. In: Proceedings of EDBT

Hadjieleftheriou M, Kollios G, Bakalov P, Tsotras VJ (2005) Complex spatio-temporal pattern queries. In: Proceedings of VLDB

Jensen CS, Lin D, Ooi BC (2004) Query and update efficient B+-tree based indexing of moving objects. In: Proceedings of VLDB

Jensen CS, Tiesyte D, Tradisauskas N (2006) The COST benchmark—comparison and evaluation of spatio-temporal indexes. In: Proceedings of DASFAA

Li F, Cheng D, Hadjieleftheriou M, Kollios G (2005) On trip planning queries in spatial databases. In: Proceedings of SSTD

Manolopoulos Y, Nanopoulos A, Papadopoulos AN, Theodoridis Y (2005) R-trees: theory and applications. Springer, New York

Mokbel MF, Aref WG (2007) Location-aware query processing and optimization (invited tutorial). In: Proceedings of MDM

Mouratidis K, Hadjieleftheriou M, Papadias D (2005) Conceptual partitioning: an efficient method for continuous nearest neighbor monitoring. In: Proceedings of SIGMOD

Mouratidis K, Yiu M, Papadias D, Mamoulis N (2006) Continuous nearest neighbor monitoring in road networks. In: Proceedings of VLDB

Myllymaki J, Kaufman J (2003) DynaMark: a benchmark for dynamic spatial indexing. In: Proceedings of MDM

Papadias D, Zhang J, Mamoulis N, Tao Y (2003) Query processing in spatial network databases. In: Proceedings of VLDB

Patel J, Yu JB, Kabra N, Tufte K, Nag B, Burger J, Hall N, Ramasamy K, Lueder R, Ellmann C, Kupsch J, Guo S, Larson J, De Witt D, Naughton J (1997) Building a scalable geo-spatial DBMS: technology, implementation, and evaluation. In: Proceedings of SIGMOD

Pelanis M, Šaltenis S, Jensen CS (2006) Indexing the past, present, and anticipated future positions of moving objects. ACM Transactions on Database Systems 31(1):255–298

Pelekis N, Kopanakis I, Marketos G, Ntoutsi I, Andrienko G, Theodoridis Y (2007) Similarity search in trajectory databases. In: Proceedings of TIME

Pfoser D, Jensen CS, Theodoridis Y (2000) Novel approaches to the indexing of moving object trajectories. In: Proceedings of VLDB

Pfoser D (2002) Indexing the trajectories of moving objects. IEEE Data Engineering Bulletin 25(2):3–9

Pfoser D, Jensen CS (2003) Indexing of network constrained moving objects. In: Proceedings of GIS

Ray S, Simion B, Brown AD (2011) Jackpine: a benchmark to evaluate spatial database performance. In: Proceedings of ICDE

Šaltenis S, Jensen CS, Leutenegger ST, Lopez MA (2000) Indexing the positions of continuously moving objects. In: Proceedings of SIGMOD

Sandu Popa I, Zeitouni K, Oria V, Barth D, Vial S (2011) Indexing in-network trajectory flows. The VLDB Journal 20(5):643–669

Shahabi C, Koladhouzan M, Sharifzadeh M (2003) A road network embedding technique for k-nearest neighbor search in moving object databases. Geoinformatica 7(3):255–273

Sharifzadeh M, Kolahdouzan M, Shahabi C (2008) The optimal sequenced route query. The VLDB Journal 17(4):765–787

Sistla AP, Wolfson O, Chamberlain S, Dao S (1997) Modeling and querying moving objects. In: Proceedings of ICDE

Song Z, Roussopoulos N (2001) K-nearest neighbor search for moving query point. In: Proceedings of SSTD

Stonebraker M, Frew J, Gardels K, Meredith J (1993) The SEQUOIA 2000 storage benchmark. In: Proceedings of SIGMOD

Tao Y, Papadias D (2001) MV3R-Tree: a spatio-temporal access method for timestamp and interval queries. In: Proceedings of VLDB

Tao Y, Papadias D, Shen Q (2002a) Continuous nearest neighbor search. In: Proceedings of VLDB

Tao Y, Papadias D, Zhang J (2002b) Cost models for overlapping and multi-version structures. ACM Transactions on Database Systems 27(3):299–342

Tao Y, Papadias D, Sun J (2003a) The TPR*-tree: an optimized spatio-temporal access method for predictive queries. In: Proceedings of VLDB

Tao Y, Sun J, Papadias D (2003b) Analysis of predictive spatio-temporal queries. ACM Transactions on Database Systems 28(4):295–336

Tao Y, Cheng R, Xiao X, Ngai WK, Kao B, Prabhakar S (2005) Indexing multi-dimensional uncertain data with arbitrary probability density functions. In: Proceedings of VLDB

Theodoridis Y, Vazirgiannis M, Sellis TK (1996) Spatio-temporal indexing for large multimedia applications. In: Proceedings of ICMCS

Theodoridis Y, Sellis TK, Papadopoulos AN, Manolopoulos Y (1998) Specifications for efficient indexing in spatio-temporal databases. In: Proceedings of SSDBM

Theodoridis Y (2003) Ten benchmark database queries for location-based services. The Computer Journal 46(6):713–725

Tiakas E, Papadopoulos AN, Nanopoulos A, Manolopoulos Y, Stojanovic D, Djordjevic-Kajan S (2009) Searching for similar trajectories in spatial networks. Journal of Systems and Software 82(5):772–788

Trajcevski G, Wolfson O, Hinrichs K, Chamberlain S (2004) Managing uncertainty in moving objects databases. ACM Transactions on Database Systems 29(3):463–507

Vieira MR, Tsotras VJ (2011) Complex motion patterns for trajectories. In: Proceedings of ICDEW

Vlachos M, Kollios G, Gunopulos D (2002) Discovering similar multidimensional trajectories. In: Proceedings of ICDE

Yu X, Mehrotra S (2003) Capturing uncertainty in spatial queries over imprecise data. In: Proceedings of DEXA

Chapter 5
Moving Object Database Engines

Spatial database research has focused on supporting the modeling and querying of geometries associated with objects in a database. Regarding static spatial data, the major commercial as well as open source database management systems (e.g. DB2, MySQL, Oracle, PostgreSQL, SQL Server) already provide appropriate data management and querying mechanisms that conform to Open Geospatial Consortium (OGC) standards. On the other hand, temporal databases have focused on extending the knowledge kept in a database about the current state of the real world to include the past, in the two senses of "the past of the real world" (*valid time*) and "the past states of the database" (*transaction time*). The recent years' effort is an attempt to achieve an appropriate kind of interaction between both sub-areas of database research. Spatiotemporal databases are the outcome of the aggregation of time and space into a single framework. As delineated in several surveys in the literature of spatiotemporal databases, a serious weakness is that the majority of the proposed approaches deals with few common characteristics found across a number of specific applications. Thus the applicability of each approach to different cases, fails on spatiotemporal behaviors not anticipated by the application used for the initial model development. For the previous reason, the field of the Moving Objects Database (MOD) has emerged and has shown that it presents the most desirable properties among various proposals. However, although a lot of research has been carried out in the field of MOD, most of the efforts do not pay attention into embedding the proposed algorithms (i.e. access methods and query processing techniques) on top of existing DBMS where real-world organizations base on. The goal of this chapter is to describe effective frameworks capable of aiding either an analyst working with mobility data, or more technically, a MOD developer in implementing a MOD in a real DBMS.

N. Pelekis and Y. Theodoridis, *Mobility Data Management and Exploration*,
DOI 10.1007/978-1-4939-0392-4_5, © Springer Science+Business Media New York 2014

5.1   From Spatial Database Systems to MOD Engines

Researchers have been challenged by an effective representation of moving objects in a database. If it is only time-dependent locations that need to be managed (mobile phone users, cars, ships, etc.), then moving point (i.e. the trajectory of a moving point object) is the basic abstraction; while, if the time-dependent shape or extent is also of interest (e.g. group of people, armies, spread of spills), then we are talking about moving regions. In the remainder, we will only discuss the case of trajectories of moving objects.

As already discussed in Chap. 3, a straightforward approach widely accepted in the literature is to model a moving point as a sequence of tuples, each of which is a location-time pair of the form $<p, t>$, indicating that the object is at position p at time t, where p in its turn is a coordinate pair (x, y). Points are stored either as spatial objects (of type point) in a spatial DBMS or as pairs of numbers in a traditional DBMS, and a database query language (e.g. SQL) is used to retrieve the location information. This method is called *point-location management*, and it has several critical drawbacks, as it does not enable interpolation between successive point, it may lead to a critical precision/resource trade-off, while it results into cumbersome and inefficient software development. The realization of these shortcomings had as a result a new line of research, where moving points are viewed as three- (2-dimensional space + 1-dimensional time) or even higher dimensional entities.

Abstract data types for objects that their movement follows the *sliced representation* presented in Chap. 3 were implemented in a *"moving objects"* extension package in suitable extensible database architectures. The prototype was developed as an algebra module for the SECONDO database system, which was the first MOD engine that appeared in the literature.

An alternative approach following the sliced representation is the Hermes MOD engine. Hermes presents an integrated and comprehensive design of moving object data types in the form of extensible modules that can be embedded in Open Geospatial Consortium (OGC) compliant Object-Relational Database Management Systems (ORDBMS), taking advantage of their extensibility interface. Hermes provides the functionality to construct a set of moving geometries, which are just variables of simple continuous functions that obtain hypostasis when projected to the spatial domain (i.e. becoming OGC spatial data types) at a specific instance in time. Each one of these moving objects is supplied with a set of methods that facilitate the user to query and analyze mobility data. Embedding this functionality offered by Hermes in an ORDBMS data manipulation language, one obtains a flexible, expressive and easy to use query language for moving objects fully embedded in real OGC-compliant ORDBMS.

In a different application domain, PLACE system presents a framework for the incremental and continuous evaluation of queries over spatiotemporal data streams, where there is shared execution among the concurrent continuous queries. Nevertheless, since the emphasis of this book is on historical mobility data management (trajectories of moving objects), in the following paragraphs we provide a few more details on the two most representative MOD developments, SECONDO and Hermes.

Fig. 5.1 The architecture
of SECONDO system

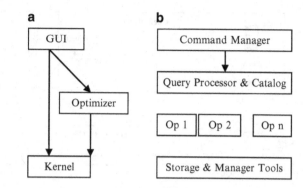

5.1.1   SECONDO

SECONDO is a DBMS prototype platform especially adjusted to be extended by algebra modules for non-standard applications. It does not support a predetermined data model, rather is open for implementation of new models. As illustrated in Fig. 5.1a, its architecture consists of (i) a *kernel*, which offers query processing over a set of implemented type system algebras, (ii) an *optimizer*, which implements the essential part of an SQL-like language, and (iii) an extensible GUI where new data types and models can provide specialized viewers for moving objects.

SECONDO objects are stored (and retrieved) by the *Storage Manager* into a database and managed by the *Catalog* (see Fig. 5.1b). More specifically, in order to realize the spatiotemporal data type model (i.e. algebra) discussed earlier, two modules have been built. The first provides (stationary) spatial data types and operations (the ROSE algebra module) and the second provides the moving object algebra module. Both modules utilize standard and relational algebras. The moving object algebra module, namely the implementation of all the time-dependent data types, uses the *sliced representation* modeling a time-dependent value as a sequence of simple temporal functions.

Having defined the physical storage of each of the objects in the type system, the next step is the development of the temporal counterparts of operations defined in the ROSE algebra. For example, an operation answering whether a point resides inside a region or not, such as

$$\text{inside}[\text{point} \times \text{region}] \rightarrow \text{bool}$$

is transformed, by an approach called *lifting*, to an operation returning a time-varying boolean representing the periods where a moving point is inside the region, such as

$$\text{inside}[\text{mpoint} \times \text{region}] \rightarrow \text{mbool}$$

Finally, special operators for moving types are offered, including projections into the temporal or spatial domain, methods that determine rate of change properties (e.g. speed of a moving object), etc. The above-described functionality has been embedded into a SQL-like query language, where queries can be written directly as PROLOG terms, as this is the development language of the optimizer.

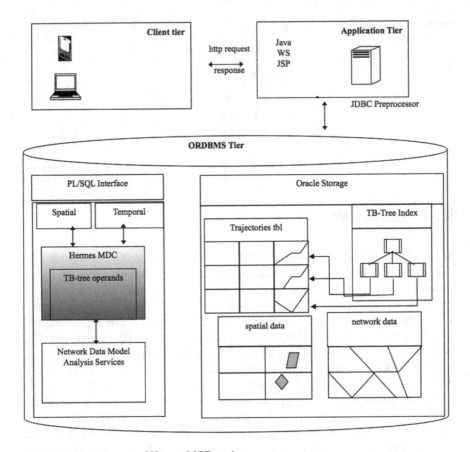

Fig. 5.2 The architecture of Hermes MOD engine

5.1.2   Hermes

Hermes aims to support modeling and querying of continuously moving objects, taking advantage of the extensibility interfaces provided by modern ORDBMS. In particular, Hermes MOD engine is developed as a system extension in two versions, providing trajectory functionality to Oracle and PostgreSQL ORDBMS, respectively. Although ultimately the two versions support more or less the same functionality based on a similar model, there are few differences. In the remainder, we will present Hermes implying the version on top of Oracle. In detail, Hermes defines a trajectory data type and a collection of operations, which is further enhanced by a special trajectory preserving access method, namely the TB-tree (presented in Chap. 3). Figure 5.2 illustrates Hermes architectural framework.

According to Fig. 5.2, Hermes resides at the ORDBMS tier. In detail, the Oracle ORDBMS Server enhanced with trajectory data storage and query capabilities serves as the infrastructure for mobility data management. In order to implement such a framework in the form of a data cartridge, Hermes exploits a set of standard data types together with (stationary) spatial data types offered by Oracle Spatial, and appropriate temporal types. Embedding this functionality in the database query language of the underlying ORDBMS, one obtains an expressive and easy way to manipulate moving objects. Accessing Hermes' API can be done exactly the same ways as any native component of the underlying ORDBMS.

5.2 A Data Type Model for Trajectory Databases

So far, we have presented a general view of MOD engines. In this section, we dig into the infrastructure of the extensible ORDBMS and describe all the necessary steps required in order to introduce a data type system for trajectories of moving objects. More specifically, we describe the base, temporal and spatial types that compose the basic constructs for the definition of the moving objects data types. The discussion complies with Hermes data type system, however the presentation is generic and applicable to any other OGC-compliant system.

5.2.1 Preliminaries of Trajectory Data Types

In order to define a data type model for Trajectory Databases (TD), we need to base on standard database types built into any DBMS, as well as temporal and spatial types.

Temporal types are responsible for providing pure temporal functionality. All existing ORDBMS provide temporal literal data types found in *ODMG* object model (namely, Date, Time, Timestamp and Interval). However, in order to support moving objects one would require additional temporal object data types, such as Period (defined by two timestamps corresponding to the starting and ending time-points of a temporal period) and Temporal_Element (defined as a sequence of non-overlapping period objects).

On the other hand, a reasonable choice to base the definition of moving objects data types on top of corresponding spatial types (point, line segment, rectangle, etc.) is to use an *OGC Geometry* (i.e. a spatial type that conforms to the specifications of the Open Geospatial Consortium). The spatial object types supported in OGIS data model are illustrated in Fig. 5.3.

Such a spatial extension is found in several state-of-the-art ORDBMS (e.g. DB2 spatial extender, MySQL spatial extension, Oracle Spatial, PostGIS, SQL Server) and provides an integrated set of functions and procedures that enable spatial data following the OGC standard to be efficiently stored, accessed, and further processed in a spatial database. Of course, the geometric operations forming the behavior of spatial types supported by these extensions, handle queries statically, meaning that *there exists no notion of time associated to the spatial objects*. This is exactly the goal addressed in the type system for trajectories that we present below.

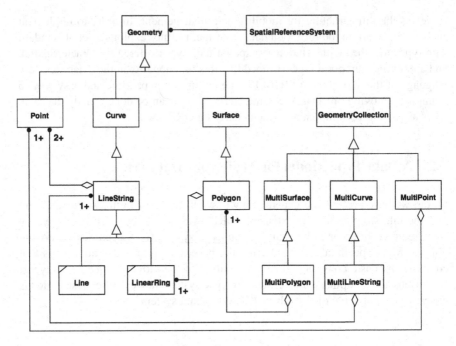

Fig. 5.3 Spatial object types in OGIS data model

5.2.2   Trajectory-Oriented Data Types

Recalling the discussion on sliced representation (Chap. 3), a moving point may be decomposed in a sequence of *unit moving points* (i.e. slices, according to the sliced representation), with each of them representing a simple movement function. As such, in order to define a unit moving point, we need to associate a period of time with the description of a simple function that models the behavior of the moving point in that specific time period.

Based on this approach, two real-world notions are directly mapped to our model as object types, namely time period and simple function. The first concept can be supported by a closed-open period data type. The second concept may be a new object type, named Unit_Function, defined as a tuple $<p_s, p_c, p_e, flag>$, where points p_s and p_e represent the starting (x_s, y_s) and ending (x_e, y_e) coordinates of the sub-motion defined, p_c corresponds to the center (x_c, y_c) of a circle upon which the object is moving (in case of arc motion), and *flag* denotes the kind of motion (whether it is constant, linear or arc) between the end points p_s and p_e (recall Fig. 3.2 in Chap. 3).

Combining time_period and Unit_Function, the primitive Unit_Moving_Point data type is defined. This is a fundamental type since it represents the smallest

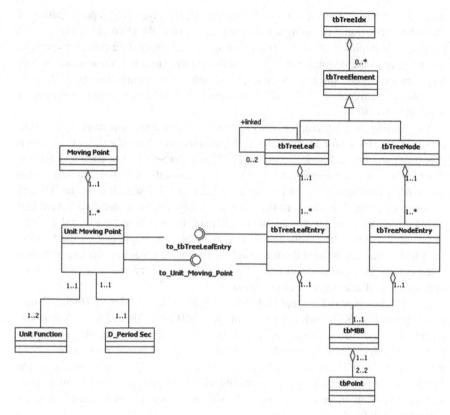

Fig. 5.4 The trajectory-oriented data type system

granule of movement of a trajectory. Based on it, one can define the moving point as a collection of unit moving points. As such, Moving_Point data type is defined over Unit_Moving_Point data type and, in terms of object orientation, there exists a composition relationship between the two.

According to the abstract data type paradigm, the trajectory type system presented above should be extended with indexing capabilities. To provide the discussion through an example, we choose the TB-tree access method presented in Chap. 4. Similar methodology can be followed if one would choose to incorporate an alternative method, such as the 3D R-tree, etc. In particular, we discuss the data types required for embedding the TB-tree in an ORDBMS that supports moving objects. We should note that these data types are transparent to the user of a MOD engine and their usage is internal, for the construction of the tree.

Figure 5.4 provides, in the form of UML class diagram, an abstract, though insightful, view of the index organization, along with the connection with the trajectory-oriented data types. The right part of the diagram includes the datatypes that are exported to the user, while the left part depicts the objects participating in the index formation. Following a top-down description, the tbTreeIdx class is used mainly for completeness as an abstraction of the corresponding part of the model and it refers to the definition of TB-tree index on the table where the actual trajectory data are stored. Since the main trajectory table may initially be empty, the corresponding aggregation with the lower-level tbTreeElement class possesses a cardinality of «0..*».

Descending the diagram, we observe that the whole arrangement is separated in two different classes for TB-tree nodes: tbTreeNode and tbTreeLeaf, for the internal (non-leaf) and the leaf nodes, respectively. Given that the size of each leaf node is fixed according to the chosen disc block size, its capacity in entries (fanout) is also fixed. As a consequence, and recalling the theoretical discussion on the TB-tree provided in Chap. 4, portions of the same trajectory may be stored in different leaf nodes, which are linked with a double linked list. This is denoted in Fig. 5.4 with the linked association. The cardinality of this association is explained by the fact the each leaf node may be linked with 0 (in case the entire trajectory fits in one node), 1 (especially for head and tail nodes, in case the trajectory spans in two or more nodes) or 2 leaf nodes (in all other cases).

An object of type tbTreeLeaf includes a number of leaf entries (tbTreeLeafEntry data type in Fig. 5.4), each consisting of the MBB (tbMBB in Fig. 5.4) that surrounds the trajectory segment kept in the leaf entry, along with an integer number (1..4) denoting its orientation. As for tbMBB, it is composed by two objects of tbPoint type, which correspond to the lower-left and upper-right corners of the MBB, while tbPoint has only a property of tbX collection type, which is an array of size 3 used to hold triplets $<t, x, y>$ of time-stamped positions. More specifically, the attributes of tbTreeLeaf class are:

- TID: the global trajectory identifier;
- ptrCurrentNode: the current node's identifier encapsulated in the object to facilitate implementation issues;
- ptrParentNode: a pointer to the parent of the current node used to ascend the tree when necessary;
- ptrPreviousNode: a pointer to the node containing the previous fragment of the same trajectory;
- ptrNextNode: a pointer to the node containing the next fragment of the same trajectory;
- LeafEntries: a fixed capacity collection of objects of type tbTreeLeafEntry, which correspond to the current leaf entries as previously described;
- count: the cardinality of LeafEntries.

In a similar fashion, an object of type tbTreeNode contains a fixed capacity collection of objects of type tbTreeNodeEntry, corresponding to each of the sub-trees. Eventually, the two interfaces, to_tbTreeLeafEntry and to_Unit_Moving_Point, illustrated in Fig. 5.4, provide essential mechanisms for object transformation between the core types of the type system.

5.3 Extending the Trajectory Data Type Model with Object Methods and Operators

In order to get real advantage of the trajectory data type model presented in the previous section, a rich functionality should be added at the SQL query language level. In this section, we present methods (functions and operations) on the data types introduced earlier. These methods are classified as follows:

- *predicates and projection methods*: operations on topological and other relationships (intersection, within distance, etc.) over a trajectory and operations that restrict a trajectory with respect to spatial and/or temporal constraints;
- *numeric operations*: functions that calculate a numeric value over a trajectory (e.g. speed);
- *distance functions:* functions that calculate distance (dissimilarity) between two trajectories of moving objects;
- *query operators:* several types of well-known queries for efficiently processing trajectory databases (timeslice, range, NN, etc.).

In the following, we describe selected operations from each class along with their functionality. All operations are overloaded in an object *mp* of type Moving_Point.

5.3.1 Predicates and Projection Methods

A MOD engine should provide a rich palette of object methods of special interest to describe, one the one hand, projections of *mp* in spatial and or temporal domain and, on the other hand, relationships between *mp* and another moving point. Examples include:

- Unit_Moving_Point *unit_type* (TimePoint *t*): among the unit moving points that compose moving point *mp*, returns the one whose time period encloses time point *t*.
- Geometry *at_instant* (TimePoint *t*): returns the geometry of the 2-dimensional point, where moving point *mp* was located at time point *t*.
- Geometry *f_initial* (): returns the geometry of the 2-dimensional point, where moving point *mp* was located at first valid time point of the trajectory's lifespan. Similarly, *f_final* returns the last location of the trajectory.
- Moving_Point *at_period* (TimePeriod *dt*): returns the portion of the trajectory of moving point *mp* during period *dt*.
- Geometry *route* (): returns the geometry of the 2-dimensional polyline that corresponds to the route followed by moving point *mp*.
- TimePoint *f_enter* (Geometry *r*): returns the timepoint when the trajectory enters the given region *r* for the first time. Similarly, there is a method (called *f_leave*) for returning the time when the object leaves the region.

Fig. 5.5 A trajectory
resampled every 5, 10,
20 and 100 timestamps
(from *left* to *right*)

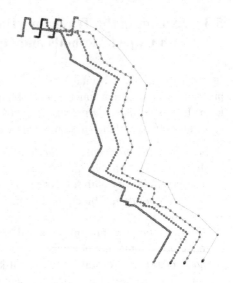

- MultiPoint *get_enter_leave_points* (Geometry *r*): returns the locations (as a
  multipoint collection geometry) in which a trajectory enters or leaves a given
  region *r*.
- Moving_Point *f_intersection* (Geometry *r*): returns the portion of the trajectory
  of moving point *mp* whose route is inside polygon region *r*.
- boolean *f_within_distance* (number *d*, Moving_Point *mp2*, TimePoint *t*): returns
  True/False with respect to whether the projections (points) of moving points *mp*
  and *mp2* at time point *t* lie within Euclidean distance *d* from each other.
- set ⟨Moving_Point⟩ *re_sample*(set ⟨Moving_Point⟩ *mps*, set ⟨TimePoint⟩ *tps*):
  the function takes as input a collection of moving points *mps* and a collection of
  timepoints and returns the moving points sampled only at the given timepoints,
  if these are part of their lifespan.

To demonstrate the effect of e.g. *re_sample* function, Fig. 5.5 illustrates a trajectory sampled every 5, 10, 20 and 100 timestamps.

As already mentioned, the above methods are representative of the Hermes API,
while there are others that exhibit similar behavior. For instance, similar to *at_
period*(), there are more projection operations, like the *at_point*() and *at_linestring*()
methods that restrict a moving point to a static point or linestring geometry, respectively, or return the temporal point or period that the restriction is valid. Moreover
there are simple SQL functions that aid the analyst to visualize a selected set of
trajectories and the query input/output, via exporting the corresponding data to files
whose format makes them directly visible by well-known GIS software tools.

Chapter 12, which is a hands-on with Hermes@Oracle, illustrates the outcomes of several SQL scripts that use the methods discussed in this section.

5.3.2 Numeric Operations

MOD engines support a class of object methods that either compute a numeric value or quantify a property related with the rate of change of a trajectory. Examples include:

- number *f_length* (TimePoint *t*): returns the length of the 2-dimensional polyline that corresponds to the route followed by moving point *mp* from the beginning of the movement till a given time point *t*.
- number *f_speed* (TimePoint *t*): returns the speed (calculated as time derivative of distance) of moving point *mp* at time point *t*.
- number *f_direction* (Moving_Point *mp2*, TimePoint *t*): returns the angle ϕ in degrees ($0 \leq \phi < 360$) formed between *xx'* axis and the line segment that connects two points, which are the projections of moving points *mp* and *mp2* at time point *t*.

It is also worth to be noted that, based on this method, two sets of directional operations may be defined: the first set consists of four operations—*f_west()*, *f_east()*, *f_north()*, and *f_south()*—that return True/False depending on whether a reference geometry is located e.g. *west* of the moving point *mp* at a specific time point; the second set consists of another four operations, namely *f_left()*, *f_right()*, *f_above()*, and *f_behind()*, which return True/False taking into account the relative position of moving point *mp* with respect to another moving point.

5.3.3 Distance Functions

Calculating (dis)similarity between trajectories of moving objects is essential in MODs. It is so, not only because of similarity search ("*given a reference trajectory, search for similar ones in the trajectory database*"—see relevant discussion in Chap. 4) but also for analysis and data mining purposes (this aspect will be discussed in detail in Chap. 6). Below, we present a palette of distance functions as object methods in the adopted trajectory data type model. The discussion that follows considers the *Locality In-between Polylines* (LIP) distance between two trajectories, which is defined over the area of the closed polygon that is formed when two polylines (i.e. projections of two trajectories in 2-dimensional space) are considered, if necessary, after connecting the two starting points and the two ending points of the polylines. Examples of distance (dissimilarity) functions include:

- number *GenLIP* (Moving_Point *mp2*): returns the LIP distance between the two 2-dimensional polylines that correspond to the routes followed by the moving points *mp* and *mp2*.
- number *GenSTLIP* (Moving_Point *mp2*): returns the LIP distance between the two 2-dimensional polylines that correspond to the routes followed by the

synchronized sub-trajectories of moving points *mp* and *mp2*, i.e. the portions of the two trajectories that co-exist in time.

– number *number_of_times_close*(Moving_Point *mp2*, number *d*, number *tolerance*): this method simply counts how many times the current moving object came close to the given one (i.e. *mp2*) in terms of an Euclidean distance threshold *d*. In order to simplify calculations, the proximity of the two trajectories is evaluated only to a predefined sequence of timestamps. As such, the two trajectories are first re-sampled (by using, internally, the previously presented *re_sample* procedure), so as for both the trajectories, we have pre-calculated the locations they reside at the same set of timepoints, which in this case is the union of timestamps of the two trajectories. Then the distance evaluation takes place for all the common timepoints.

5.3.4 Query Operators

Given the discussion in Chap. 4 about various queries of interest in MODs, here we present a set of operators that implement some of the most popular ones, which could be either coordinate-based (*range*—note that *timeslice* is a special case of *range* query, *point NN*, and *trajectory NN*) or trajectory-based (*topological*), according to the trajectory data model we have presented:

– set ⟨Moving_Point⟩ *Range* (Geometry *r*, TimePeriod *dt*): returns the sub-trajectories that conform to both spatial (*r*) and temporal (*dt*) constraints.
– set ⟨number, Unit_Moving_Point⟩ *IncrPointNN* (Geometry *p*, TimePeriod *dt*, number *k*): returns *k* pairs of moving object identifiers and their respective trajectories (actually their 3-dimensional segments), according to their closeness with respect to point *p* during time period *dt*.
– set ⟨number, Unit_Moving_Point⟩ *IncrTrajectoryNN* (number *id*, number *k*): returns *k* pairs of moving object identifiers and their respective segments, according to their similarity with the trajectory of moving point identified by *id*; similarity between trajectories is calculated according to the *Minimum Horizontal Euclidean* distance function.
– set ⟨number⟩ *Topological* (Geometry *r*, TimePeriod *dt*, String *mask*): returns the identifiers of trajectories that obey a topological relationship (according to the value of *mask*: "Enter", "Leave", "Cross", etc.) with (rectangular) region *r* during time period *dt*.

These operators perform queries on a MOD table that stores trajectories indexed by an indexing technique (TB-tree), implementing the respective queries provided in Chap. 4. In Chap. 12, the reader will have the chance for a hands-on experience with a real MOD engine, Hermes@Oracle, which fully complies with the previously described trajectory data type model.

5.4 On Mobility Data Provenance

Although there exist significant developments compared to the youth of the field of moving object databases, there is a lack of methods and tools that would enable researchers and practitioners working in the field have their one-stop place where they could find available datasets, implementations of algorithms, respective metadata, case studies and other useful material about mobility data. *Scientific research is considered of good provenance when it is documented in detail sufficient to allow reproducibility.* To support this, also for the field of mobility data management and mining, we should be able to keep the lineage not only of the data and the algorithms used/devised so far, but also their metadata, and, more important, the workflows (e.g. sequences of queries and analysis tasks) documented in the literature or in empirical case studies that advance the domain. Such metadata and scientific workflows assist scientists and programmers with tracking their data and techniques through their interpretations, analyses, and transformations.

Related work in the context of the above-discussed concepts includes approaches for the provenance of traditional kind of data. Not going into the details of these approaches, none of these may be used as-is on mobility data, as they are tailored for legacy data instead. Most of them use workflows to perform scientific experiments and simulations. In general, provenance should be associated not solely with data, but, more important, with the processes that lead to their creation.

Regarding mobility data provenance, a recent effort is the so-called ChoroChronos. org geo-portal. ChoroChronos.org can be considered as a hybrid approach in that, both workflows and MOD queries are supported (the latter is achieved by the connection with Hermes MOD engine). Regarding user functionality, ChoroChronos. org offers a palette of options through its interface (Fig. 5.6). Aided by Hermes, users are allowed to filter the MOD content according to spatial and/or temporal criteria on the stored trajectories as well as criteria on metadata items. The (either original or filtered) datasets can be visualized on top of maps and extracted to various formats for own use. ChoroChronos.org users may attach comments on the existing datasets and algorithms. They are also encouraged to donate their own content in the portal through a user-friendly interface.

Regarding the content itself, ChoroChronos.org supports the submission of datasets of moving objects and algorithms from registered users who also post associated metadata, such as source, thumbnail visualization, information about attributes, data type (GPS, GSM, WiFi etc.), aimed task (clustering, classification, etc.), scientific area (e.g. wildlife animal behavior), relevant papers and citations, etc. Alternative access options ("accessible to all"; "accessible to registered users/ groups only"; "meta-data only available") make the donation procedure flexible enough to conform to different requirements, for example, projects that have produced mobility datasets or algorithms and have to obey specific publishing or distribution rules due to intellectual property rights.

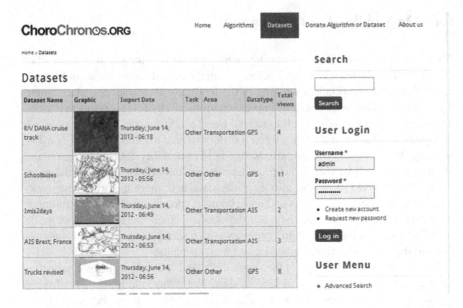

Fig. 5.6 The interface of ChoroChronos.org geo-portal

Another useful concept provided by ChoroChronos.org is the support of workflows and processes and the "inheritance" or "semantic linkage" of datasets or algorithms. In particular, assuming that the outcome of a process (database query or data mining algorithm) on a dataset is another dataset, a user can semantically link the two datasets by providing the necessary information required for the reproducibility of this analysis. The same functionality is supported for algorithms; for example, a query processing or clustering algorithm may be later improved to require less parameters or to be applicable for other purposes too. As such, ChoroChronos.org supports user-defined semantic Directed Acyclic Graphs (DAG), which is the tool to support the workflows required for data provenance. Browsing datasets and algorithms along these semantics DAG is also supported, so as the user may find the part of the DAG's hierarchy of interest.

5.5 Summary

In this chapter we reviewed a few MOD frameworks and, more important, we studied how such a framework may be designed and developed, by following a step-by-step methodology that includes development of data types in a ORDBMS and implementation of methods (operations and functions) on these data types. The methodology we presented focuses on handling space and time properties of moving objects as "first class citizens" at the database level, implying the foundation of a datatype-oriented model for moving objects and its development inside a DBMS by taking into advantage of its extensibility interfaces. Mobility data provenance is also an important aspect. We outlined the (admittedly, few found in the literature) data repository approaches focusing on MODs.

Following this line, one may enumerate many interesting extensions related to issues already discussed in the previous chapters. For instance, it is interesting to interrelate advanced trajectory reconstruction techniques to MODs. It is interesting to investigate how could we reconstruct trajectories from other than GPS recordings, such GSM, WiFi, RFID, etc. types of data, which of course have their own peculiarities. The same stands also for map-matching algorithms, which should be integrated to MODs. In other words, it is important to put into the game, the networks whereupon objects are moving, as in majority, objects are moving according to the restrictions of a network (i.e. a road network for vehicles, or a virtual but predetermined network for vessels).

Regarding trajectory query processing, although in the previous chapter we studied a lot of interesting types of queries, there should be made an effort of realizing these on top of real MOD engines and their query languages, instead of ad hoc implementations which are hard or impossible to be repeated in extensible DBMSs. A nice selection of such a category of queries is advanced mobility mining techniques, which will be discussed in Chap. 7. Moreover, being able to support dynamic, real-time services implies that MODs should be able to support hybrid (past- and present-locations) indexing techniques, which is another open issue. Last but not least, we should mention that mobility data management in the past two decades has mainly focused on centralized MOD architectures. The vast interest of users for novel, dynamic applications, in the era of big, linked data, for sure enforces the redesign of MOD architectures, following new computing paradigms such as Map-Reduce, column-based, etc. A relevant discussion will follow in Chap. 10.

5.6 Exercises

Ex. 5.1. What is the difference between the map-matching operation (presented in Chap. 3) and the *at_linestring* method mentioned in Sect. 5.3.1?

Ex. 5.2. Try to sketch an algorithm for the *f_intersection* function. Is there a difference if the geometry input parameter is a rectangle or polygon (possibly with holes)?

Ex. 5.3. The query operators presented in Sect. 5.3.4 are index-based, which means that the trajectories stored in the database are indexed using e.g. TB-trees and indexed-based search is performed instead of sequential search in the table storing trajectories. Having this in mind, sketch an algorithm for the *Range* operator using the TB-tree index.

Ex. 5.4. Assume a courier company, whose vehicles are equipped with GPS devices transmitting their space-time location to a central MOD. In order for the company to manage and analyze the motion of its fleet, you are asked to:

(a) Design a database, consisting of a table for the trajectories of the vehicles (by using data types as those introduced in this chapter), a table for the road network (representing edges of the road network as linestring geometries), and a table for landmarks on the road network (representing points of interest, such as petrol stations, etc., as point geometries).

(b) Feed the database with a few realistic data.

(c) Apply a fleet management system scenario to the database, in the form of the following set of queries (you are asked to provide the respective SQL statements and query results):

   (i)   Find all vehicles following (or crossing) highway 'H'.

   (ii) Find all vehicles moving inside a given region and time period.

   (iii) If vehicle 'X' is in the answer set of (ii), when and where did it enter the region?

   (iv) What distance did vehicle 'X' travel inside the region?

   (v) Give a list of options to the driver of vehicle 'X' to refuel the vehicle in a distance less than 2 km from its current (= the most recently recorded) location.

Ex. 5.5. Access ChoroChronos.org geo-portal. Browse among the available datasets; choose one of your preference and "play" with it by posing queries via the available interface or by using your custom tools. When you complete your analysis, donate the new dataset that corresponds to the result of your analysis to ChoroChronos.org. Link the two datasets by providing the necessary metadata that allow the reproducibility of your analysis.

5.7 Bibliographical Notes

Simulating trajectory data types in spatial extensions of commercial DBMS and, hence, providing indexing support in MODs, has been exercised in Oracle Spatial (Kothuri and Ravada 2002) and in PostGIS (Giannotti et al. 2011). The Oracle approach exploits on the *sdo_geometry* data type and linear referencing system (LRS) functions operating on the geometry data. As for indexing, spatial and temporal dimensions are indexed separately in a 2-dimensional and a 1-dimensional R-tree, respectively, as discussed in Kothuri and Ravada (2002). On the other hand, PostGIS supports 2-, 3-, and 4-dimensional geometry data types; hence, the trajectory of a moving object could be simulated by a 3-dimensional polyline. Of course, in this case the peculiarity of the time dimension is not taken into consideration.

Regarding MOD prototype engines, the research line with respect to SECONDO was initiated by the model proposed in Güting et al. (2000), Erwig and Schneider (2002). Then, Forlizzi et al. (2000) provided an initial study of algorithms for a rather large set of operations, while Lema et al. (2003) presented a comprehensive, systematic study of algorithms for a subset of these operations. This work also offered a blueprint for implementing such a *"moving objects"* extension package for suitable extensible database architectures. The outcome of this work was demonstrated in de Almeida et al. (2006). The stream-based MOD engine PLACE is presented in Mokbel and Aref (2005). The Hermes MOD engine is presented in

Pelekis et al. (2006, 2008, 2014). The set of methods for similarity search between trajectories of moving points as they have been discussed in Sect. 5.3.4, are presented in Pelekis et al. (2012).

In the literature on provenance, one may find extensive surveys, like Simmhan et al. (2005) and Bose and Frew (2005) that propose a taxonomy of provenance techniques. In the spatial domain, Lanter (1991) discusses derived data in GIS and characterizes lineage as the information describing transformations to derive the data. Greenwood et al. (2003) extends Lanter's view as the metadata recording the process of experiment workflows, annotations, and notes about experiments. The statement about the good provenance of scientific research quoted in Sect. 5.4 is due to Düntgen et al. (2009). The data provenance line of research is pursued by past and current initiatives showing the increasing importance of data provenance. Examples of such initiatives are the EU-funded GridProvenance.org and the NSF-funded DataONE.org and DataConservancy.org. ChoroChronos.org and MoveBank.org provide geo-portal services. Especially for ChoroChronos.org, the interested reader may find more details in Pelekis et al. (2011). Its name (meaning, *space and time*, in Greek) origins from a pioneering European basic research project that put the foundations, at least at European level, of MOD research (Koubarakis et al. 2003).

References

Bose R, Frew J (2005) Lineage retrieval for scientific data processing: a survey. ACM Comput Surv 37:1–28

de Almeida VT, Güting RH, Behr T (2006) Querying moving objects in secondo. In: Proceedings of MDM

Düntgen C, Behr T, Güting RH (2009) BerlinMOD: a benchmark for moving object databases. VLDB J 18(6):1335–1368

Erwig M, Schneider M (2002) Spatio-temporal predicates. IEEE Trans Knowl Data Eng 14(4):881–901

Forlizzi L, Güting RH, Nardelli E, Schneider M (2000) A data model and data structures for moving objects databases. In: Proceedings of SIGMOD

Giannotti F, Nanni M, Pedreschi D, Pinelli F, Renso C, Rinzivillo S, Trasarti R (2011) Unveiling the complexity of human mobility by querying and mining massive trajectory data. VLDB J 20(5):695–719

Greenwood M, Goble C, Stevens R, Zhao J, Addis M, Marvin D, Moreau L, Oinn T (2003) Provenance of e-science experiments—experience from bioinformatics. In: Proceedings of AHM

Güting RH, Bohlen MH, Erwig M, Jensen CS, Lorentzos NA, Schneider M, Vazirgiannis M (2000) A foundation for representing and querying moving objects. ACM Trans Database Syst 25(1):1–42

Kothuri RKV, Ravada S (2002) Spatio-temporal indexing in Oracle: issues and challenges. IEEE Data Eng Bull 25(2):56–60

Koubarakis M, Sellis T et al (eds) (2003) Spatio-temporal databases: the chorochronos approach. Springer, Berlin

Lanter DP (1991) Design of a lineage-based meta-data base for GIS. Cartogr Geogr Inf Syst 18(4):255–261

Lema JAC, Forlizzi L, Güting RH, Nardelli E, Schneider M (2003) Algorithms for moving objects databases. Comput J 46(6):680–712

Mokbel MF, Aref WG (2005) PLACE: a scalable location-aware database server for spatio-temporal data streams. IEEE Data Eng Bull 28(3):3–10

Pelekis N, Theodoridis Y, Vosinakis S, Panayiotopoulos T (2006) Hermes—a framework for location-based data management. In: Proceedings of EDBT

Pelekis N, Frentzos E, Giatrakos N, Theodoridis Y (2008) Hermes: aggregative LBS via a trajectory DB engine. In: Proceedings of SIGMOD

Pelekis N, Stefanakis E, Kopanakis I, Zotali C, Vodas M, Theodoridis Y (2011) ChoroChronos. org: a geoportal for movement data and processes. In: Proceedings of COSIT

Pelekis N, Andrienko G, Andrienko N, Kopanakis I, Marketos G, Theodoridis Y (2012) Visually exploring movement data via similarity-based analysis. J Intell Inf Syst 38(2):343–391

Pelekis N, Frentzos E, Giatrakos N, Theodoridis Y (2014) HERMES: A Trajectory DB engine for mobility-centric applications. Int J Knowl Based Org 4(1) (in press)

Simmhan YL, Plale B, Gannon D (2005) A survey of data provenance in e-science. SIGMOD Rec 34(3):31–36

Part III
Mobility Data Exploration

The only source of knowledge is experience.

Albert Einstein

Chapter 6
Preparing for Mobility Data Exploration

Exploring mobility data already collected and stored in efficient database systems is the next step of mobility data management. Historical data 'hide' a treasure of 'buried' knowledge that 'asks' for mining. To do so, the typical Knowledge Discovery in Data (KDD) process typically includes the organization of historical information in a Data Warehouse (DW), a first level of analysis exploiting on data cubes build upon the DW, according to a multi-dimensional model, and, then, a deeper look into the data in order to extract models and patterns that data obey or follow, using data mining techniques. In this chapter, we provide the preparatory actions in order for data mining to follow in the next chapter. In particular, we present DW approaches for mobility, especially for trajectory data, and we discuss about the kind of multi-dimensional analysis that is suitable for mobility data and the challenges that arise due to its peculiarity. We also provide sound definitions for trajectory similarity, which is a key component of whatever analysis to be made with trajectory databases.

6.1 Mobility Data Warehousing

Data Warehousing (DW) has been widely investigated for conventional (non-spatial) and, recently, for spatial data. The key concept of DW is the *data cube*, a view over the database stored in DW, with its typical structure (called, *star schema*) consisting of a *fact table* with measures to be analyzed and a number of *dimension tables* that determine the dimensions of the analysis; recall the example in Fig. 2.9b.

Assuming a MOD to be the operational database that feeds the DW, we result in the so-called *Trajectory DW* (TDW). In this case, a star schema for TDW would consist of (at least) a temporal and a spatial dimension, along with thematic, application-oriented dimension(s). On the other hand, the measures maintained in the fact table should reflect information about movement to be analyzed. For instance, interesting TDW measures could include the number of trajectories, their (average/minimum/maximum) distance traveled, the (average/minimum/maximum) duration of their trip,

N. Pelekis and Y. Theodoridis, *Mobility Data Management and Exploration*, 121
DOI 10.1007/978-1-4939-0392-4_6, © Springer Science+Business Media New York 2014

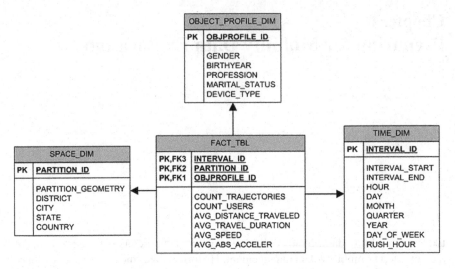

Fig. 6.1 An example data cube for trajectories (Marketos et al. 2008)

aggregations over derived movement information (e.g. speed, acceleration, heading, etc.), representative trajectory(-ies), etc.

6.1.1   Modeling Trajectory Data Cubes

An example data cube for trajectory data is illustrated in Fig. 6.1. In this example, the data cube consists of three dimensions and six measures to be analyzed. In particular:

- *Dimensions* include space, time and object profile along with their hierarchies (tables SPACE_DIM, TIME_DIM, and OBJECT_PROFILE_DIM, respectively).
- *Measures* include aggregate information over trajectories that correspond to the specific <space, time, object profile> triple, namely: COUNT_TRAJECTORIES (the number of distinct trajectories), COUNT_USERS (the number of distinct users), AVG_DISTANCE_TRAVELED, AVG_TRAVEL_DURATION, AVG_ SPEED, and AVG_ABS_ACCELER (for those trajectories, the average distance traveled, the average duration of the trip, the average speed, and the average acceleration (in absolute values), respectively).

Of course, choosing a reasonable resolution in space/time is critical. As usual, there is tradeoff between quality and usage of resources: the finer the resolution the better the quality of analysis (but, also, the larger the database). This can be made clear through the example of Fig. 6.1. Let us suppose that (a) the resolution in space and time dimension is 1 % of the total extent per coordinate and 10 min, respectively, (b) there exist ten different user profiles, (c) the history of DW goes 3 years back, and (d) the refresh of the DW is performed every week. Regarding the dimension tables, these values result in 10,000 (=100 × 100 cells) records in table SPACE_DIM, ~150,000 (=3 × 365 × 24 × 6) records in table TIME_DIM, which are extended by ~1,000 (=7 × 24 × 6) records after each refresh, and 1,000 records in table OBJECT_PROFILE_DIM. In the most extreme case, the 3-years fact table would contain about 1.5 trillion records, extended by about ten billion records every week!! Fortunately, the fact table is very sparse (it is quite probable that many combinations of space partitioning, time partitioning and user profile find zero support in the source trajectory database), so it can be manageable.

It is clear that useful analysis can be performed over trajectory data, and TDW is the required infrastructure in order to perform it. However, in order to maintain such a TDW over a trajectory database (stored e.g. in a MOD such as the ones discussed in Chap. 4), some challenging issues arise:

- during ETL process: how to efficiently feed the fact table? Since MOD is a complex database, specialized algorithms are expected to appear here.
- during OLAP process: how to effectively roll-up over the dimension hierarchies? The so-called *"distinct count problem"* is introduced here.

Both issues are of great significance for a smooth TDW architecture. Efficient ETL process guarantees fast communication between MOD and TDW during refreshing the aggregated data stored in TDW. On the other hand, OLAP should be effective since the quality of the analysis results is essential for decision-making.

6.1.2 Performing ETL Process

ETL process in data warehouses consists of *extracting* source data, *transforming* if and when necessary, and *loading* transformed, aggregate information into the data warehouse repository, so as this data to be reflected in the data cubes. Considering the TDW paradigm (e.g. Fig. 6.1), feeding the dimension tables is, in general, straightforward. As soon as the resolution is space and time dimensions has been set (the size of space partitioning in table SPACE_DIM and the length of time intervals in table TIME_DIM, respectively), fresh information in the source MOD is segmented and reflected in the respective dimension tables.

On the other hand, feeding the fact table, i.e. calculating its measures with respect to the new entries inserted into the dimension tables is not straightforward at all. Technically, in order to calculate the measures, we have to extract the portions of the trajectories that fit into the base cells of the data cube. In other words, for every combination of space partitioning, time partitioning and user profile (assuming the example of Fig. 6.1) we have to select the trajectories (actually, the sub-trajectories) that support it.

Leaving user profile out of discussion since this is a filter on the thematic properties of a trajectory, we concentrate on space and time partitioning and how this can filter our trajectory database. For instance, a trajectory that spans over three zones in space partitioning and two intervals in time partitioning may be segmented to

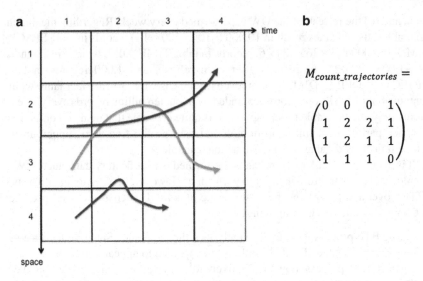

Fig. 6.2 Exemplifying the ETL process in TDW: (**a**) two trajectories; (**b**) the respective COUNT_TRAJECTORIES measure

(at least) three up to (at most) six sub-trajectories during ETL. All these sub-trajectories should be analyzed (for their length, duration, speed, acceleration, etc.) in order for the measures of the fact table to be calculated.

As an example, three trajectories are depicted in Fig. 6.2a (for visualization purposes, space is illustrated in a single axis), visiting four to six (spatiotemporal) base cells each, out of a total of sixteen cells. Focusing on cell (2, 3), two sub-trajectories of the red and blue trajectory are identified and their movement features are calculated in order for the respective measures to be calculated; e.g. the respective COUNT_TRAJECTORIES measure is illustrated on the right (Fig. 6.2b).

Having, on the one hand, the spatiotemporal filters (space and time partitioning) and, on the other hand, the trajectory database, the ETL problem is formalized as follows:

The ETL problem in Trajectory DW: *Given a trajectory database T consisting of a set of trajectories T_i, a space partitioning S consisting of a set of space partitions (spatial cells) s_j, and a time partitioning τ consisting of a set of time partitions (temporal cells) τ_k, the ETL problem is to segment each T_i in one or more sub-trajectories, each of which is spatially and temporally contained in s_j and τ_k, respectively, which, in turn, will contribute to filling the measure matrix M of size $|S| \times |\tau|$.*

There exist at least two alternative solutions to this problem that arise naturally: the *cell-oriented* and the *trajectory-oriented* approach. In the former, for every cell in the spatiotemporal partitioning, the trajectory database is queried in order for the respective sub-trajectories to be identified (and contribute to measure calculations). In the latter, for every trajectory in the database, the involved cells in the spatiotemporal partitioning are identified, and their measures are updated. Formally, the respective algorithms are provided below.

Algorithm Cell-oriented-ETL-approach (COA)

Input: trajectory database T, space partitioning S, time partitioning τ

Output: measure matrix M

1. FOR each (spatiotemporal) base cell $c_{jk} = (s_j, \tau_k)$ in S $\times$ τ
2. Search for sub-trajectories in T that are contained in c_{jk}
3. Compute measures M[j,k]

Algorithm Trajectory-oriented-ETL-approach (TOA)

Input: trajectory database T, space partitioning S, time partitioning τ

Output: measure matrix M

1. FOR each trajectory T_i in T
2. Find the (spatiotemporal) base cell $c_{jk} = (s_j, \tau_k)$ in S $\times$ τ, T_i is contained in
3. Compute measures M[j,k]

Whether COA or TOA is more efficient depends on several parameters, including the size of the trajectory database, the space/time resolution adopted, etc. For sure, COA can be extremely fast if its expensive step, searching the trajectory database (Line 2), is index supported in MOD.

6.2 OLAP Analysis in Trajectory Data Cubes

As already discussed, having filled the fact table with appropriate measures, we can perform effective OLAP analysis (in terms of roll-up, drill-down, slice and cross-over operations).

Here is an example of progressive OLAP analysis for the purposes of e.g. a car insurance company using the TDW scheme of Fig. 6.1:

- *"Display the number of users and their average speed, for each space partition and per hour"*, this requires working at the hour level in time dimension (roll-up operation in table TIME_DIM);
- *"In the above result, focus on downtown area, night hours and young drivers, and display their average speed"*, this requires working at the district level in space dimension (roll-up operation in table SPACE_DIM) and selecting downtown area (slice operation in table SPACE_DIM), night hours (slice operation in table TIME_DIM) and young drivers (slice operation in table OBJECT_PROFILE_DIM);
- *"In the above result, retrieve those users that are 'responsible' for average speed over the speed limit and check whether (when and where) they exceeded this speed limit"*, this requires looking back to the feeding trajectory database (cross-over operation).

The aggregations at higher levels of hierarchy (roll-up operations) are essential for effective multi-dimensional analysis like the above. However, a problem that arises is that *a trajectory may contribute to several (spatiotemporal) base cells*. This side effect should be taken into consideration when rolling-up. This is the so-called "*distinct count problem*".

6.2.1 Addressing the Distinct Count Problem

In order to present the distinct count problem through an example, let us recall Fig. 6.2. Assuming that we perform a roll-up operation on the downtown area, which is the union of the four centrally located cells, (2,2), (2,3), (3,2), and (3,3), the question is whether the *count* measures, such as COUNT_TRAJECTORIES, can be simply summed up, as in conventional data warehousing, or not. Clearly, the answer is no. Summing up in this example would result to a value 7, which is far from reality (the correct value is, of course, 3). The reason for this deviation is that every trajectory may be counted multiple times since it may contribute to multiple base cells. The distinct count problem is formalized as follows:

The distinct count problem during aggregating trajectory data cubes: *Given a space partitioning S consisting of a set of space partitions (spatial cells) s_j, a time partitioning τ consisting of a set of time partitions (temporal cells) τ_k, and a measure matrix M of size $|S| \times |\tau|$ including count measures over trajectories, the distinct count problem is to estimate as better as possible the resulting measure after aggregating in space and time due to a roll-up operation.*

A suboptimal solution to the problem proposes that we keep a note on the border between two base cells, in our case, the number of trajectories that have crossed the border. By summing up these inter-cell 'cross' values and subtracting them from the original sum of counts, we get closer to the actual aggregate value. Following this approach, in the example of Fig. 6.2a, the number of trajectories that crossed the border between cells (2,2) and (2,3) is two (the blue and the red trajectory), and so on. Aggregating cells (2,2), (2,3), (3,2) and (3,3), we initially calculate a sum equal to 7, and then we subtract 4 (=2+1+0+1, the 'cross' values in the four borders). So, we get the correct result, 3!

Although successful in our example, this approach is suboptimal because it cannot guarantee that it will always find the correct aggregate value. When trajectories follow strange paths, the 'cross' values cannot detect all problematic cases and the result may deviate from the correct.

6.2.2 Indexing Summary Information for Efficient OLAP

Efficient OLAP processing requires aggregate data to be stored in appropriate indexes (as detailed data is stored in B+-trees, R-trees, etc. for efficient query processing at the operational level). Of course, summarized results can be obtained

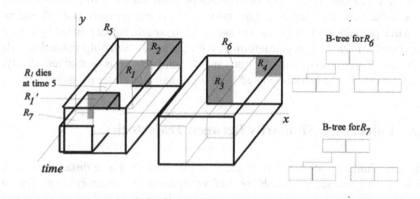

Fig. 6.3 The a3DRB-tree for indexing spatiotemporal aggregate data (*source*: Tao and Papadias 2005)

using conventional operations on individual objects but this may not our choice for several reasons: in some cases, detailed data may not be accessible from an OLAP server due to privacy issues, or even they may be unavailable due to the excessive volumes which would be required to store them.

If we consider, at the finest aggregation level, a given space and time partitioning, then each region is associated with a set of measures, which are usually numeric (e.g. number of moving objects, average speed) or, in advanced cases complex (e.g. representative trajectory). On such a measure, we may perform a *window aggregate query*, which returns the aggregate measure of the (spatiotemporal) regions intersecting a given query region Q_R during a given query interval Q_T. In order to support this type of aggregate queries for numeric measures, the a3DRB-tree has been proposed.

a3DRB-tree. The concept followed by a3DRB-tree is that a 3D R-tree index hosts the spatiotemporal regions formed at the finest aggregation level and its entries are associated with respective B^+-tree-like structures (aggregate B-trees) that store information about the numeric measures corresponding to each region; this scheme is illustrated in Fig. 6.3. In particular, an entry r of the a3DRB-tree has the form <r.MBR, r.ptr, r.aggr, r.lifespan, r.btree>, where r.MBR and r.ptr are defined as in a normal 3D R-tree, r.aggr keeps the aggregated measure about r over r.lifespan, and r.btree points to an aggregate B-tree, which stores the detailed measure information of r at concrete timestamps. Regions may remain static or may change; in the example of Fig. 6.3, region R_1 may change to R_1' at timestamp 5, hence creating a new covering node R_7.

Given a query region Q_R and a query interval Q_T, processing of window aggregate queries using a3DRB-trees is performed as follows: the algorithms starts from the root of the 3DR-tree, and for each entry r, one of the following conditions holds: either r is covered by both Q_R and Q_T (case i) or r is covered by Q_R and partially overlaps Q_T (case ii) or r partially overlaps Q_R and partially overlaps (or is inside)

Q_T (case iii) or none of the previous conditions holds (case iv). In case i, the good news is that pre-computed aggregate information is already there (r.aggr) and no extra tree propagation is required; in case ii, the aggregate B-tree pointed by r.btree is accessed to retrieve aggregate information for Q_T; in case iii, the algorithm descends to the next R-tree level (pointed by r.ptr) and the same process is applied recursively; finally, in case iv, the entry is ignored since it is out of interest for the query.

6.3　Calculating Similarity Between Trajectories

The (dis-)similarity between two trajectories that is defined as a distance measure finds lots of applications in mobility data management. In trajectory simplification (see Chap. 3), it is a measure of quality between the original and the approximated trajectory; in trajectory management (see Chap. 4), similarity search is a very interesting query type; in data warehousing, the representative among a set of trajectories could be an advanced data cube measure (see Sect. 6.2). Moreover, as expected, trajectory similarity finds an ideal field for application in data mining, where almost every e.g. clustering algorithm relies on such a distance function to provide groups of similar objects and outliers (to be discussed in Chap. 7). Also, anonymization of mobility data due to privacy issues exploits on trajectory similarity (to be discussed in Chap. 8). For the above reasons, in this section we overview alternative proposals for calculating the similarity between two trajectories of moving objects and discuss their pros and cons.

6.3.1　Functions Computed over the Sampled Points

A straightforward solution in order to calculate the dissimilarity between two trajectories defined as sequences of sampled two-dimensional points is to use an L_p norm. Such a distance metric for two vectors R and S is calculated by the formula:

$$L_p\left(\vec{R}-\vec{S}\right)=\left(\sum_{i=1}^{n}|r_i-s_i|^p\right)^{\frac{1}{p}}$$

L_p metrics have several advantages, such as linear computation cost, they are easy to implement, they require no special parameters, and they obey the triangle inequality (as such, they can be used by well-known indices applicable to metric spaces). However, their main disadvantage in order to be applicable to mobility data, is that they require absolute temporal synchronization of the two trajectories, therefore they are very sensitive to noise, and not applicable when there is a time shift between the lifespans of the two trajectories or when they have different lengths (i.e. number of points). To eliminate this limitation a solution is to use linear interpolation when sampling times of intermediate points do not match, or when the two tracks have different number of points. This is illustrated in Fig. 6.4a where the two trajectories, which have the same starting and ending timestamps, are interpolated at all timestamps when one of the two or both trajectories have samples.

Fig. 6.4 (a) Euclidean distance and (b) average trajectory between two trajectories

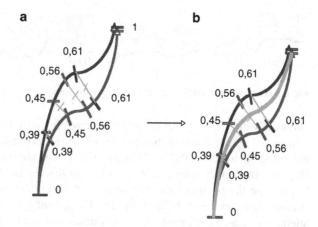

As such, the Euclidean distance L_2 norm for two trajectories of moving objects, R and S, is calculated by the following equation after transforming them to six-dimensional vectors (i.e. $R(r_1, r_2, \ldots, r_n)$, $S(s_1, s_2, \ldots, s_n)$, denoted by the vertical segments upon the two paths).

$$L_2(R,S) = \sqrt{\sum_{i=1}^{n}\left(r_{i,x} - s_{i,x}\right)^2 + \left(r_{i,y} - s_{i,y}\right)^2}$$

Note that this approach for transforming the trajectories of the moving objects to vectors in a multi-dimensional space is only one of different possible ways to follow. Subsequently, we will examine another way for such a transformation, which, as we will see in Chap. 7, is a mandatory step for many data mining algorithms aiming at discovering interesting mobility patterns. Figure 6.4b illustrates a simple example that computes the "average" trajectory of the two trajectories in question, which can be thought of as a representative trajectory of the collective behavior of the initial two.

A drawback of the Euclidean distance is that by computing the square deviations, makes outliers have a great impact on the overall distance. An alternative to Euclidean distance which handles outlier values more effectively, is the Manhattan distance, which uses the sum of the absolute differences of the coordinates (i.e. setting parameter $p=1$ to the equation of L_p norm). In the same line but on the other extreme, the Chebyshev distance does not use all the absolute differences of the coordinates of the two trajectories, but only the largest one, resulting to the following Equation:

$$L_\infty(R,S) = \max_{i=1}^{n}\left|r_{i,x} - s_{i,x}\right| + \left|r_{i,y} - s_{i,y}\right|$$

The *Dynamic Time Warping* (DTW) distance function, which has been originally used in speech recognition, can also be used for calculating the similarity between two trajectories which differ in length or speed. The key feature of DTW is that it

Fig. 6.5 Euclidean vs. DTW distance function in timeseries

allows a sequence to "stretch" or to "shrink" in order to better fit with another sequence. Figure 6.5 shows the mapping of the points of two timeseries when using either the Euclidean or the DTW function. Although the two timeseries are similar, they are not aligned in time, having as a result that the Euclidean distance matches one-by-one the points of the two series and falsely concludes that there is large distance between them. Instead, the DTW does not have this problem because it attempts to match each point of the first timeseries with the most appropriate ones of the second and finally chooses the shortest distance.

The DTW method for calculating the distance between two trajectories of moving objects, R and S, is given by the following equation:

$$DTW(R,S) = L_p(r_n, s_m) + \min \begin{Bmatrix} DTW(R, Head(S)), \\ DTW(Head(R), S), \\ DTW(Head(R), Head(S)) \end{Bmatrix}$$

where $Head(R) = ((r_{1,x}, r_{1,y}) \ldots (r_{n-1,x}, r_{n-1,y}))$ and L_p may be any L metric.

The *Longest Common SubSequence* (LCSS) is another similarity measure that tries to match two timeseries allowing their "stretching", however, without changing the sequence of elements, while allowing some elements of the sequences to be left unmatched. The main advantage with respect to the previously presented techniques, is that the LCSS handles noise more efficiently because it can disregard noisy points (by not necessarily matching them), in contrast to other techniques that have to match all points (even the noisy ones). The LCSS uses two parameters; the parameter δ indicating the temporal range wherein the method searches to match a specific point, and the ε parameter which is a distance threshold to indicate whether two points match or not. Formally, the LCSS between two trajectories of moving objects, R and S, is given by the following equation:

$$LCSS(R,S) = \begin{cases} 0, & m = 0 \, or \, n = 0 \\ LCSS(Head(R), Head(S)) + 1, & if \, |r_{n,x} - s_{m,x}| < \varepsilon \\ & and \, |r_{n,y} - s_{m,y}| < \varepsilon \\ & and \, |n - m| \leq \delta \\ \max\{LCSS(Head(R), S), LCSS(R, Head(S))\}, & otherwise \end{cases}$$

where $Head(R) = ((r_{1,x}, r_{1,y}) \ldots (r_{n-1,x}, r_{n-1,y}))$, δ is an integer and ε is a real number.

Having computed the similarity between two trajectories, the distance between them is calculated by the following equation which normalizes the similarity value with respect to the length of the compared trajectories:

$$D_{\delta,\varepsilon}(R,S) = 1 - \frac{LCSS_{\delta,\varepsilon}(R,S)}{\min(n,m)}$$

Figure 6.6 presents the above two similarity functions in action. In particular, Fig. 6.6a depicts the application of DTW in two trajectories whereas Fig. 6.6b demonstrates the idea of LCSS parameters δ and ε; intuitively, the points of the two trajectories that lie within the gray area (determined by the parameters δ and ε) can be matched.

The *Edit Distance on Real Sequences* (EDR) function is based on the well-known edit distance function which has been successfully used in applications like bioinformatics where an important task is to quantify the similarity between two strings. Given two strings, the Edit Distance function calculates the minimum number of insertions, deletions and replacements in order for the two strings to become identical. Like LCSS, EDR also uses a threshold ε to match two points, however the distance may be either 0 or 1, depending on whether they match or not. Applying this matching criterion to two points r_i and s_j of trajectories of moving objects:

$$match(r_i,s_j) = \begin{cases} 1, & if\ |r_{i,x} - s_{j,x}| \le \varepsilon\ and\ |r_{i,y} - s_{j,y}| \le \varepsilon \\ 0, & else \end{cases}$$

Due to this choice, outliers have much less impact on the overall distance, in comparison to the Euclidean distance or the DTW. On the other hand, a crucial difference with LCSS, is that EDR adds a penalty for the gaps between two matched parts, which is proportional to the gap length. This leads to higher accuracy in calculating the distance, in contrast to LCSS. Given the above, the EDR for two trajectories of moving objects, R and S, with lengths n and m, respectively, is defined as:

$$EDR(R,S) = \begin{cases} n, & if\ m = 0 \\ m, & if\ n = 0 \\ \min \begin{cases} EDR(Rest(R),Rest(S)) + subcost, \\ EDR(Rest(R),S) + 1, \\ ERP(R,Rest(S)) + 1 \end{cases}, & otherwise \end{cases}$$

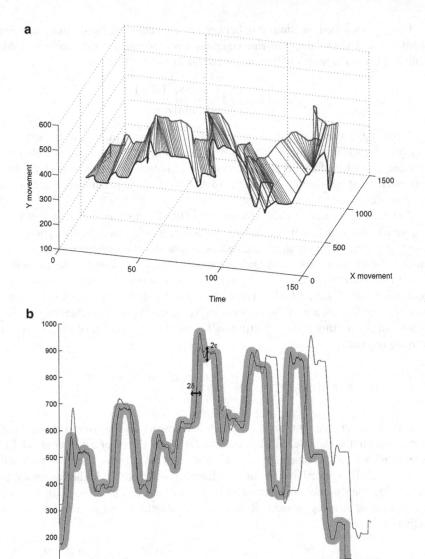

Fig. 6.6 Time-stretching dissimilarity calculation between trajectories: (**a**) DTW distance function (*source*: Vlachos et al. 2006); (**b**) LCSS similarity function (*source*: Vlachos et al. 2002b)

where $Rest(R) = ((r_{2,x}, r_{2,y}) \dots (r_{n,x}, r_{n,y}))$ and the *subcost* is defined as:

$$subcost = \begin{cases} 0, & if\ match(r_1, s_1) = 1 \\ 1, & otherwise \end{cases}$$

The functions presented so far can be classified into two broad categories: metric functions, which can not handle the time shift of the trajectories (e.g. L_p), and non-metric functions that support time shift (e.g. DTW, LCSS, EDR). The *Edit Distance with Real Penalty* (ERP) function combines the advantages of the two categories. Furthermore, ERP tries to combine the advantages of DTW and EDR by using a fixed reference point for calculating the distance of the points that have been unmatched. Moreover, if the distance between the two points is very large, ERP uses the distance between a point and the reference point. ERP is defined by the following equation:

$$
ERP(R,S) = \begin{cases} \sum_{i=1}^{n} dist(s_i, g), & \text{if } m = 0 \\ \sum_{i=1}^{n} dist(r_i, g), & \text{if } n = 0 \\ \min \begin{cases} ERP(Rest(R), Rest(S)) + dist(t_1, s_1), \\ ERP(Rest(R), S) + dist(r_1, g), \\ ERP(R, Rest(S)) + dist(s_1, g) \end{cases}, & \text{otherwise} \end{cases}
$$

Last but not least, *Swale* belongs to the same category like LCSS and EDR that use a distance threshold in order to match two points. Moreover, when Swale calculates the total distance, it penalizes the case when two points do not match (gap cost), while it rewards the case when a matching takes place (match reward). The evaluation of Swale demonstrated improved accuracy in comparison to DTW, ERP, LCSS and EDR, however, its main disadvantage is that it requires setting two additional parameters (i.e. the gap cost and the match reward). Swale is defined in the following equation:

$$
Swale(R,S) = \begin{cases} n * gap_c, & \text{if } m = 0 \\ m * gap_c, & \text{if } n = 0 \\ reward_m + Swale(Rest(R), Rest(S)), & \text{if } |r_{1,x} - s_{1,x}| \le \varepsilon \\ & \text{and } |r_{1,y} - s_{1,y}| \le \varepsilon \\ \max \{ gap_c + Swale(Rest(R), S), gap_c + Swale(R, Rest(S)) \}, & \\ & \text{otherwise} \end{cases}
$$

where gap_c is the gap cost and $reward_m$ is the match reward parameters.

6.3.2 Computing the Similarity Between Entire Trajectories or Sub-trajectories

The distance functions presented in the previous section calculate the distance between two trajectories according to their sampled points only. On the other hand, a generalization of the previously discussed Euclidean distance is the DISSIM function that defines the dissimilarity between two trajectories being valid during the period $[t_1, t_n]$ as the definite integral of the function of time of the Euclidean distance between the two trajectories during the same period:

$$DISSIM(R,S) = \int_{t_1}^{t_n} L_2(R(t), S(t)) dt$$

In order to simplify the computation of this function, since both trajectories can be represented by the union of their sampled points using linear interpolation, the above definition of dissimilarity is transformed to:

$$DISSIM(R,S) = \sum_{k=1}^{n-1} \int_{t_k}^{t_{k+1}} L_2(R(t), S(t)) dt$$

where t_k are the timestamps when moving objects R and S recorded their position.

The previous transformation actually decomposes the dissimilarity calculation problem to computing a series of integrals, each of which is upon two three-dimensional segments, i.e. two points moving with linear functions of time between consecutive timestamps t_k and t_{k+1}. This function still includes complex computations. A solution for this is to approximate the computation of the integrals by using the trapezoid rule. The following equation gives approximated computation of the previous one:

$$DISSIM(R,S) \approx \frac{1}{2} \sum_{k=1}^{n-1} \left(\left(L_2(R(t_k), S(t_k)) + L_2(R(t_{k+1}), S(t_{k+1})) \right) \cdot (t_{k+1} - t_k) \right)$$

The distance functions presented so far, aggregate (i.e. sum) some measures that are defined upon the constituents of trajectories, which are three-dimensional points or segments (depending on the adopted data model). Instead of calculating the (e.g. Euclidean) distance of the projections of two trajectories at the same timestamp, or its generalization that is applied on pairs of segments, an alternative approach handles time as a third spatial dimension and defines the distance between three-dimensional line segments that compose the trajectory as the minimum energy of transportation of segment r_i to segment s_j. This idea has been introduced on *Earth Movers Distance* (EMD) and has been successfully applied on pattern recognition and computer vision applications. In our case, this energy can be defined by the sum of two energies: (i) translation energy $d_\perp(r_i, s_j)$ and (ii) rotation energy $d_\angle(r_i, s_j)$; the former depends on the Euclidean distance while the latter depends on the angle formed between the two segments.

Fig. 6.7 The TRACLUS
approach for calculating
the distance between two
directed segments
(*source*: Lee et al. 2007)

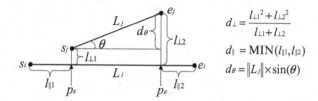

The distance $d_\perp(r_i, s_j)$ can be estimated using the following definition of distance between two points $p_r = (x_r, y_r, t_r)$ and $p_s = (x_s, y_s, t_s)$ of r_i and s_j, respectively, that have the minimal distance in the 3D space. The spatial and temporal differences of the two points are made comparable by using two weights, w_1 and w_2, for space and time, respectively.

$$d \perp (r_i, s_j) = \sqrt{w_1 \cdot (x_r - x_s)^2 + w_1 \cdot (y_r - y_s)^2 - w_2 \cdot (t_r - t_s)^2}$$

Weight w_1 may be provided by the user. Then, w_2 can be automatically estimated by the ratio w_2/w_1. It holds that the ratio w_2 determines the spatial difference (e.g. how many meters) corresponds to one time unit difference (e.g. 1 s).

On the other hand, $d_\angle(r_i, s_j)$ expresses the distance that a point residing on the line segment (of the minimum length) will cover, after its rotation of θ rads, to meet the other (longer) segment. Let $|r_i|$ and $|s_j|$ denote the lengths of the two three-dimensional line segments, respectively. Then, $d_\angle(r_i, s_j)$ is defined as follows:

$$d\angle (r_i, s_j) = \min (|r_i|, |s_j|) \cdot |\theta|$$

Both $d_\perp(r_i, s_j)$ and $d_\angle(r_i, s_j)$ are expressed in Euclidean distance units in order for the resulted value to be comparable with each other.

A similar approach, used in TRACLUS framework for trajectory clustering purposes (to be discussed in detail in Chap. 7), quantifies the distance between *directed segments*, which actually correspond to sub-trajectories. More specifically, three component functions are defined between two directed segments, L_i and L_j: (i) perpendicular $d_\perp$, (ii) parallel $d_\parallel$, and (iii) angular $d_\angle$ distance (illustrated in Fig. 6.7). The overall distance is simply their weighted sum, where different weights may be used according to different application requirements.

Focusing on the projection of trajectories in the spatial plane, two moving objects may be considered similar when they move close (i.e. their routes approximate each other) at the same place, irrespective of time. The *Locality In-between Polylines* (LIP) defines such a distance function upon the (projected on the Cartesian plane) routes of the trajectories. The idea is to calculate the area of the shape formed in-between the two two-dimensional polylines that correspond to the routes of the two trajectories. An example is illustrated in Fig. 6.8, where the five shaded areas are the ones that contribute in LIP. Graphically, the area between two polylines is the one traversed by the one polyline when it appropriately moves, shrinks and/or extents itself towards

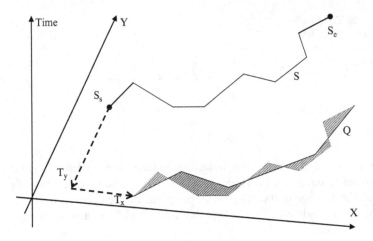

Fig. 6.8 The Locality In-between Polylines (LIP) distance function

the other polyline so as to perfectly match each other. Interestingly, one should note that such a distance measure is meaningful when the two trajectories move (more or less) towards the same direction, implying that this procedure is followed after the partitioning of the two trajectories in sub-trajectories that fulfil this property.

Another approach for computing the distance between two trajectories, which also demonstrates a different technique to transform trajectories to point data in a vector space, is to approximate each trajectory in a MOD with a sequence of regions wherefrom the trajectory has passed. The cardinality of such a sequence may be user defined and exploits on a partitioning of the time axis into equi-durations intervals (e.g. 1 min periods). Actually, this cardinality implies the dimensionality of the produced vector space. Regarding the spatial dimension, the approach assumes a tessellation of the space with a regular grid of equal rectangular cells with user-defined size (e.g. 100×100 m^2). Given this setting, and inspired by the Piecewise Aggregate Approximation (PAA) technique used in the timeseries domain, a trajectory having n points is partitioned into $p \ll n$ equi-sized temporal periods and substitutes the three-dimensional line segments of each period with the set of the grid cells that the trajectory crosses during this period. Note that, depending on the choice of the spatial and temporal granularity, a trajectory may introduce *gaps* (i.e. regions with empty set of cells due to the fact that there is no motion during the particular period of time).

For instance, assuming a target dimension $p = 5$, the trajectory illustrated in Fig. 6.9 by the black line, augmented to form a buffer area around it (e.g. due to GPS imprecision), is transformed to five (so-called) *uncertain regions*, each of which consists of the cells differently coloured in Fig. 6.9. Given this, the distance between two uncertain regions can be defined as the minimum Euclidean distance between the MBRs of the regions or any set-theoretic distance function, upon the crossed cells of the trajectories.

Fig. 6.9 Transforming a
trajectory into a
p-dimensional vector of
regions

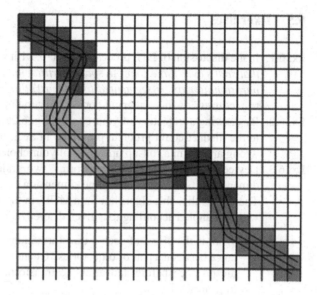

Having surveyed a number of trajectory (dis-)similarity functions, from an
analyst point of view, one may run a variety of analysis tasks, such as:

- (Task 1) *"Find groups of trajectories that follow similar routes (i.e. projections
  of trajectories on two-dimensional plane) during the same time interval (e.g.
  co-location and co-existence from 3 pm to 6 pm)"*; this is *spatiotemporal similarity*;
- (Task 2) *"Find groups of trajectories by taking only their route into consideration (i.e. irrespective of time)"*; this is *(time-relaxed) spatial-only similarity*;
- (Task 3) *"Find groups of trajectories that follow a given direction pattern (e.g.
  first NE and then W)"*; this is *trajectory derivative-based similarity*;
- and so on.

As such, it is a real challenge that we take into consideration various mobility
factors characterizing a trajectory (locality, temporality, directionality, rate of
change, etc.) and formulate a flexible framework that is able to compare the members of a trajectory database based on the above factors.

6.4 Summary

In this chapter we studied two vertical issues in mobility data management that may
be considered as stepping stones upon which knowledge discovery tasks over
mobility data may base. In detail, we show how earlier experience in multi-
dimensional analysis can be applied to trajectories by designing data cubes tailored
to analyse trajectories and to tackle their peculiarities as the distinct count problem.
Moreover, we explored several ways to quantify the (dis-)similarity between trajectories that is a key component of several data mining techniques, which is exactly
the subject of the subsequent chapter.

6.5   Exercises

Ex. 6.1. In the discussion of ETL process in Sect. 6.1.1 it was argued: *"a trajectory that spans over three zones in space partitioning and two intervals in time partitioning may be segmented to (at least) three up to (at most) six sub-trajectories"*. Is this argument correct? Try to prove its correctness or not.

Ex. 6.2. By drawing a trajectory of your own in Fig. 6.2a, try to cheat the solution to the distinct count problem presented in Sect. 6.1.2, and make it give a wrong answer in a roll-up operation of your choice.

Ex. 6.3. Device an algorithm that transforms two trajectories into a higher-dimensional space w.r.t. the time dimension as exemplified in Fig. 6.4. Then generalize this algorithm for a MOD, in the sense that all trajectories will be sampled according to a given discretization of the temporal axis.

    (a) What is the effect of such a transformation when calculating the distance between a pair of trajectories? Quantify this effect by computing the distance before and after the transformation for a few pairs of trajectories.

    (b) Could this transformation affect the calculation of measures in a trajectory data warehouse?

    (c) Recall to further study this effect in the subsequent chapter by experimenting with off-the-self hierarchical clustering algorithms that require the calculation of distance matrices for all pairs of trajectories in a MOD.

Ex. 6.4. How would you adapt the distance functions initially used in the timeseries domain, such as the DWT, LCSS, EDR, ERP etc., so as to be applicable and effective either to sequences of three-dimensional segments or to sequences of regions wherefrom the trajectories have passed?

6.6   Bibliographical Notes

Efficient schemes for trajectory data warehousing are proposed in Orlando et al. (2007), Marketos et al. (2008), Leonardi et al. (2014). The distinct count problem discussed in Sect. 6.1.2 was defined and first addressed in Tao et al. (2004). The suboptimal solution to the distinct count problem was proposed in Orlando et al. (2007). Efficient methods for OLAP analysis on trajectory data warehouses are presented in Leonardi et al. (2010), Marketos and Theodoridis (2010).

    The window aggregate query for spatial data is studied in Papadias et al. (2002). Its extension for spatiotemporal data warehouses is studied in Tao and Papadias (2005), where the authors propose the a3DRB-tree presented in Sect. 6.3. Other related work in spatiotemporal data aggregation includes Tao et al. (2004) where spatiotemporal aggregation using sketches is proposed.

As already discussed, research around trajectory similarity is inspired by the time series analysis domain. The first proposal following the paradigm of mapping a trajectory into a vector in a feature space and using an L_p-norm distance function was proposed by Agrawal et al. (1993) who adopt the Discrete Fourier Transformation (DFT) to be the feature extraction technique since DFT preserves the Euclidean distance. Lee et al. (2000) compute the distance between two multidimensional sequences by finding the distance between minimum bounding rectangles. Two approaches, which also use Euclidean distances, include the lower bound techniques (Chen and Ng 2004) and the shape-based similarity query (Yanagisawa et al. 2003).

DTW (Berndt and Clifford 1996) has been adopted in order to measure distances between two trajectories that have been represented as sequences of angle and arc-length pairs (Vlachos et al. 2002a). Lin and Su (2005) introduced the "One Way Distance" (OWD) function, which is shown to outperform DTW regarding precision and performance. The Longest Common Sub Sequence (LCSS) similarity measure (Bollobas et al. 2001) was used by Vlachos et al. (2002b) to define similarity measures for trajectories. Chen et al. (2005) proposed the Edit Distance on Real sequences (EDR), based on the Edit Distance (ED). Their experimental evaluation shows that the proposed distance function is more robust than Euclidean distance, DTW, and ERP, presented earlier in Chen and Ng (2004) and more accurate than LCSS, especially when dealing with trajectories having Gaussian noise. Vlachos et al. (2006) exploited on DTW and LCSS to efficiently index multi-dimensional time-series. The Swale distance function that is similar to LCSS and EDR is presented in Morse and Patel (2007).

Frentzos et al. (2007) proposed the DISSIM similarity metric and algorithms for index-based similarity search in Rtree-like indices, while Trajcevski et al. (2007) proposed an optimal and an approximate matching algorithm between trajectories, both under translations and rotations. Pelekis et al. (2012) proposed various geometric distance functions focusing on various movement parameters, such as the direction, the speed and the acceleration of trajectories.

The work by Lee et al. (2007) defines a distance function on directed segments, based on their perpendicular, parallel, and angle distance. In Panagiotakis et al. (2009), the Earth Movers Distance was adapted to be applied to three-dimensional segments of moving object trajectories. Pelekis et al. (2011) discusses ways to quantify the similarity between sequences of regions by proposing a variant of the ERP distance function. Tiakas et al. (2009) and Kellaris et al. (2013) proposed similarity functions for trajectories constrained in spatial networks.

In Ding et al. (2008) a comprehensive evaluation study of most of the above-discussed similarity measures was conducted, unfortunately on timeseries data, not including mobility data. As such, the absolute time semantics are left out of the evaluation benchmark. The reproduction of the same kind of experiments on different real mobility datasets is an interesting work that would provide insightful knowledge to the mobility data mining that is studied in the following chapter.

References

Agrawal R, Faloutsos C, Swami A (1993) Efficient similarity search in sequence databases. In: Proceedings of FODO

Berndt J, Clifford J (1996) Finding patterns in time series: a dynamic programming approach. In: Advances in knowledge discovery and data mining. AAAI/MIT Press, London

Bollobas B, Das G, Gunopulos D, Mannila H (2001) Time-series similarity problems and well-separated geometric sets. Nord J Comput 8(4):409–423

Chen L, Ng RT (2004) On the marriage of lp-norms and edit distance. In: Proceedings of VLDB

Chen L, Ozsu MT, Oria V (2005) Robust and fast similarity search for moving object trajectories. In: Proceedings of SIGMOD

Ding H, Trajcevski G, Scheuermann P, Wang X, Keogh E (2008) Querying and mining of time series data: experimental comparison of representations and distance measures. In: Proceedings of VLDB

Frentzos E, Gratsias K, Theodoridis Y (2007) Index-based most similar trajectory search. In: Proceedings of ICDE

Kellaris G, Pelekis N, Theodoridis Y (2013) Map-matched trajectory compression. J Syst Softw 86(6):1566–1579

Lee SL, Chun SJ, Kim DH, Lee JH, Chung CW (2000) Similarity search for multidimensional data sequences. In: Proceedings of ICDE

Lee JG, Han J, Whang KY (2007) Trajectory clustering: a partition-and-group framework. In: Proceedings of SIGMOD

Leonardi L, Marketos G, Frentzos E, Giatrakos N, Orlando S, Pelekis N, Raffaetà A, Roncato A, Silvestri C, Theodoridis Y (2010) T-warehouse: visual OLAP analysis on trajectory data. In: Proceedings of ICDE

Leonardi L, Orlando S, Raffaetà A, Roncato A, Silvestri C, Andrienko G, Andrienko N (2014) A general framework for trajectory data warehousing and visual OLAP. Geoinformatica, 18: (in press)

Lin B, Su J (2005) Shapes based trajectory queries for moving objects. In: Proceedings of ACM-GIS

Marketos G, Theodoridis Y (2010) Ad-hoc OLAP on trajectory data. In: Proceedings of MDM

Marketos G, Frentzos E, Ntoutsi I, Pelekis N, Raffaetà A, Theodoridis Y (2008) Building real world trajectory warehouses. In: Proceedings of MobiDE

Morse MD, Patel JM (2007) An efficient and accurate method for evaluating time series similarity. In: Proceedings of SIGMOD

Orlando S, Orsini R, Raffaetà A, Roncato A, Silvestri C (2007) Trajectory data warehouses: design and implementation issues. J Comput Sci Eng 1(2):211–232

Panagiotakis C, Pelekis N, Kopanakis I (2009) Trajectory voting and classification based on spatiotemporal similarity in moving object databases. In: Proceedings of IDA

Papadias D, Tao Y, Kalnis P, Zhang J (2002) Indexing spatio-temporal data warehouses. In: Proceedings of ICDE

Pelekis N, Kopanakis I, Kotsifakos E, Frentzos E, Theodoridis Y (2011) Clustering uncertain trajectories. Knowl Inf Syst 28(1):117–147

Pelekis N, Andrienko G, Andrienko N, Kopanakis I, Marketos G, Theodoridis Y (2012) Visually exploring movement data via similarity-based analysis. J Intell Inf Syst 38(2):343–391

Tao Y, Papadias D (2005) Historical spatio-temporal aggregation. ACM Trans Inf Syst 23(1):61–102

Tao Y, Kollios G, Considine J, Li F, Papadias D (2004) Spatio-temporal aggregation using sketches. In: Proceedings of ICDE

Tiakas E, Papadopoulos AN, Nanopoulos A, Manolopoulos Y, Stojanovic D, Djordjevic-Kajan S (2009) Searching for similar trajectories in spatial networks. J Syst Softw 82(5):772–788

Trajcevski G, Ding H, Scheuermann P, Tamassia R, Vaccaro D (2007) Dynamics-aware similarity of moving objects trajectories. In: Proceedings of ACM-GIS

Vlachos M, Gunopulos D, Das G (2002a) Rotation invariant distance measures for trajectories. In: Proceedings of SIGKDD

Vlachos M, Gunopulos D, Kollios G (2002b) Discovering similar multidimensional trajectories. In: Proceedings of ICDE

Vlachos M, Hadjieleftheriou M, Gunopulos D, Keogh E (2006) Indexing multidimensional time-series. VLDB J 15(1):1–20

Yanagisawa Y, Akahani J, Satoh T (2003) Shape-based similarity query for trajectory of mobile objects. In: Proceedings of MDM

Chapter 7
Mobility Data Mining and Knowledge Discovery

Knowledge discovery in trajectory databases is full of success stories in discovering interesting behavioral patterns of moving objects that can be exploited in several fields. Example domains include traffic engineering, climatology, social anthropology and zoology, implying application of the various mining techniques in vehicle position data, hurricane track data, human and animal movement data, respectively. Mobility data mining can be categorized according to the underlying mining methods used to discover the various collective behavioral patterns. Following this categorization method, there have been proposed works that try to identify various types of clusters of moving objects. Some methods group trajectories by considering the whole lifespan of the moving objects, while others try to identify local patterns that are valid only for a portion of their lifespan. Another line of research, which is parallel to that of clustering, focuses on representing a dataset of trajectories via an appropriate small set of objects, which are either artificial (i.e. the representatives or centroid trajectories of the clusters), or selected from the dataset itself (i.e. by some sampling methodology). Although clustering-oriented approaches prevail in the literature, there are many other interesting techniques that exhibit semantically rich mobility patterns and make the domain active in many areas of knowledge discovery. Among them, in this chapter we discuss sequential trajectory patterns discovery, classification and outlier detection techniques. The problem of predicting the future location of the moving objects has also been tackled and presented interesting results.

7.1 Clustering in Mobility Data

7.1.1 Extending Off-the-Shelf Algorithms for Trajectory Clustering

Clustering of trajectories implies partitioning of a MOD into clusters (groups), so that each cluster contains similar trajectories according to some distance measure. The majority of clustering literature has mostly focused on clustering of point

N. Pelekis and Y. Theodoridis, *Mobility Data Management and Exploration*,
DOI 10.1007/978-1-4939-0392-4_7, © Springer Science+Business Media New York 2014

(i.e. vector) data that trajectories do not conform to. This means that well-known clustering algorithms (e.g. K-means) cannot be directly applied. In order for such a partitioning algorithm to be applicable to trajectories, the latter should first be transformed to vectors in a multi-dimensional space by following one of the possible ways described in Chap. 6. Although this way the problem is overpassed, the problem of dimensionality curse arises, as trajectories will be probably transformed to high dimensional objects, otherwise the loss of information would be significant.

As such, the idea of measuring the similarity between two trajectories is an attractive solution that has been utilized as the mean to cluster trajectories. Two trajectories may be considered *similar* in a number of different aspects they may fully or partly coincide in space, or just have similar shapes, or have common start and/or end points; they may be fully or partly synchronous, or they may be disjoint in time but with similar dynamic behaviour (speed, acceleration, etc.). It depends on the application and goal of analysis, which of these aspects are relevant. As discussed in Chap. 6, many approaches have been introduced in the literature trying to quantify the (dis)similarity between trajectories, utilizing various trajectory features. Once we choose the appropriate distance function for the particular analysis task, we could compute the distance matrix for all the pairs of moving objects in hand and, in turn, apply e.g. a hierarchical clustering algorithm to group the trajectories in clusters.

On the same line, with an interesting kind of analysis, called *progressive clustering*, an analyst may obtain clusters by means of one of the distance functions and then reapply the same or different clustering algorithm to a selected cluster or a few clusters using another distance function (or different parameter settings), going deeper and deeper in her analysis. Below, we overview some of the most indicative clustering techniques found in the mobility data mining literature.

CenTR-I-FCM. The approach of transforming trajectories to vectors for clustering has been followed by the *CenTR-I-FCM* algorithm in connection with a soft clustering algorithm based on fuzzy logic, namely Fuzzy-C-Means (FCM). FCM is similar to K-means but it further considers uncertainty by allowing each data element to belong to different clusters by a certain degree of membership. Considering that input vector values are subject to uncertainty due to imprecise measurements, noise or sampling errors, which is the case of mobility data, the distances that determine the membership of a point to a cluster are also subject to uncertainty. Therefore, the possibility of erroneous membership assignments in the clustering process is evident.

CenTR-I-FCM is a three-step approach that has been proposed to deal with uncertainty in MOD and its effect on trajectory clustering. The approach initially adopts a symbolic representation and model trajectories as sequences of regions (i.e. wherefrom a moving object passes—see Chap. 6) accompanied with *intuitionistic fuzzy values*, i.e. elements of a so-called intuitionistic fuzzy set. *Intuitionistic Fuzzy Sets* (IFS) are generalized fuzzy sets that can be useful in coping with the *hesitancy* originating from imprecise information. IFS elements are characterized by two values representing, respectively, their *belongingness* and *non-belongingness* to this set. In the case of MOD where this set is the region that a trajectory possibly crosses, the above values represent the probabilities of presence and non-presence in the area. In order to exploit this information, the approach defines a novel variant

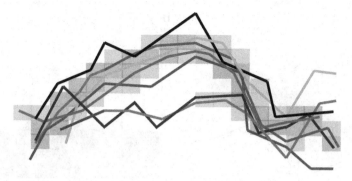

Fig. 7.1 The Centroid Trajectory (CenTra) of a set of trajectories

of the ERP distance metric, especially designed to operate on such intuitionistic fuzzy vectors, having as goal to incorporate it in some variant of the FCM algorithm that will effectively cluster trajectories under uncertainty.

The success of any FCM-based algorithm depends on the way that cluster centroids are driven towards the correct direction at each iteration of the algorithm. This direction, at each step of the algorithm, is actually decided with the help of a similarity function. However, in the MOD setting where trajectories are complex objects of different lengths, varying sampling rates, different speeds, possible outliers and different scaling factors, even the most efficient similarity function would most probably fail in different applications. Instead of using *global* similarity functions between whole trajectories, *CenTR-I-FCM* proposes to exploit *local* similarity properties between portions of the trajectories. Based on this idea, at the second step of the approach, a novel density- as well as similarity-based algorithm, called *Centroid Trajectory* (*CenTra*), is proposed to tackle the problem of discovering the centroid of a group of trajectories. Among the advantages of *CenTra* is its ability to represent complex time-aware mobility patterns that demonstrate in an intuitive way the *growth* of the pattern in space-time. The *CenTra* pattern of a set of trajectories is illustrated in Fig. 7.1: the solid lines (in different colors) correspond to trajectories while their *CenTra* is denoted by a set of coloured cells.

The background idea of the algorithm is to perform a kind of time-focused local clustering using a region growing technique under similarity and density constraints. For each time period, the algorithm determines an initial seed region (that corresponds to the sub-trajectory restricted inside the period) and searches for the maximum region that is composed of all sub-trajectories that are similar with respect to a distance threshold d and dense with respect to a density threshold δ. The seed region is determined as the one with the minimum average distance from the rest candidate regions, and which is also dense. Subsequently, the growing process begins and the algorithm tries to find the next region to extend among the most similar sub-trajectories (recall the *k-MST* type of similarity search mentioned in Chap. 4), as someone would expect to find the *best region* in one of these regions. Then, the algorithm searches among the candidate regions, i.e. those that satisfy the similarity and density constraints, in order to find the best, i.e. the one that

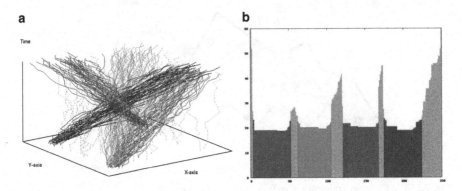

Fig. 7.2 T-OPTICS on a synthetic trajectory dataset: (**a**) four clusters of trajectories and a set of noisy trajectories; (**b**) the respective reachability plot

maximizes the average density after growing. The whole process continues until no more growing can be applied, appending in each repetition the temporally local centroid to the overall *CenTra* pattern.

At its third step, *CenTR-I-FCM* utilizes *CenTra* in its centroid update step, uses a global uncertainty-supporting similarity function to group trajectories at a higher level, and iteratively refines the results using local similarity between sub-trajectories. This algorithm has the efficiency advantages of partitioning clustering algorithms (in comparison to the higher processing cost of density-based algorithms), whereas it succeeds producing non-spherical clusters due to the inclusion of *CenTra*, that recognises representative movements of any shape.

Except for clustering uncertain trajectories, the implicit outcome of the above-described approach, namely the centroid trajectories for each of the resulted clusters, has very useful applications in several fields that could gain insight from such mobility patterns. Consider for example the domain of visual analytics on movement data. In this field it is meaningless to visualize even very small datasets, as the human eye cannot distinguish every movement pattern due to the immense size of the data. As such, more intuitive representations are required. Aiming at supporting even higher-level analysis tasks the so-called *TX-CenTra* algorithm improves the representation of *CenTra* by relaxing its point vector representation. Its idea is to identify the maximal time periods wherein the mobility pattern presents uniform behavior.

T-Optics. The idea of defining a similarity function and using it to group trajectories has also been utilized in a density-based clustering algorithm, namely the OPTICS algorithm discussed in Chap. 2. In the so-called Trajectory-OPTICS (T-OPTICS), trajectories are clustered by means of a variant of the DISSIM distance function (discussed in Chap. 6). Actually, this variant simply normalizes the integral of the DISSIM distance function with the duration of the common lifespan of the trajectories. As the algorithm is driven by the OPTICS method, the outcome of T-OPTICS is also a reachability plot. Figure 7.2a illustrates the application of T-OPTICS in a synthetic trajectory dataset consisting of four clusters and an additional set of noisy trajectories, all moving during the same period of time. The respective

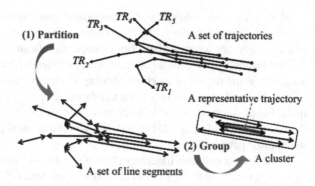

Fig. 7.3 The TRACLUS sub-trajectory clustering framework (*source*: Lee et al. 2007)

reachability plot appears in Fig. 7.2b, where the coloring of its valleys (i.e. clusters) implies successful cluster discovery.

Performing global trajectory clustering (i.e. the grouping takes into account the whole trajectories) may sometimes lead to a misleading outcome, as in the case of partial membership of objects to several clusters. *Time-Focused OPTICS* (TF-OPTICS) is actually an extension of T-OPTICS, which focuses on the discovery of the temporal intervals that lead to best clustering results. The previously mentioned temporal intervals are given by the user, so TF-OPTICS reapplies T-OPTICS on portions of trajectories, that all live in exactly the same temporal period. Then, applying some qualitative measures chooses the best clustering. The idea of identifying patterns (i.e. clusters) for a portion of the lifespan of the trajectories motivated the researchers. In the following two sections we study appropriate methods to succeed this.

7.1.2   Sub-trajectory Clustering Methods

There also exist methods whose goal is to discover local patterns created by portions of trajectories. These methods can be further classified to techniques that take the road network into account (e.g. FlowScan, NEAT) or not (e.g. TRACLUS).

TRACLUS. Trajectory Clustering (TRACLUS) is one of the first approaches that introduced a clustering algorithm aiming at identifying sub-trajectories with common behavior. The main idea of the algorithm is that trajectories are partitioned to directed segments (i.e. sub-trajectories) and then grouped irrespective of whether the trajectories as a whole are split into different clusters or not. Interestingly, after the grouping and for each cluster, a virtual representative trajectory is synthesized in order to describe the overall movement of the grouped set of directed segments. In Fig. 7.3, the *TRACLUS* partitioning and grouping framework is illustrated, where the initial five trajectories (that obviously share a common behavior only for a portion of their whole movement) are partitioned to thirteen directed segments and from them a cluster of five segments is detected, which are compactly represented by one representative trajectory.

More specifically, the *TRACLUS* workflow consists of three steps. In the first step (trajectory partitioning), the algorithm aims at discovering characteristic points of each trajectory where its behavior changes significantly. The partitioning is performed balancing two contradicting factors: *preciseness* (i.e. the difference between a trajectory and the set of its corresponding trajectory partitions should be as small as possible) and *conciseness* (i.e. the number of trajectory partitions should be minimal). Based on the *Minimum Description Length* (MDL) principle, the algorithm defines the optimal trade-off between the aforementioned properties thus performs the optimal partitioning.

In the second step (sub-trajectory clustering), the algorithm groups sub-trajectories by using a variant of the DBSCAN algorithm (presented in Chap. 2), applicable to the directed segments. A cluster consists of a density-connected set of line segments. In this particular case, the ε-neighborhood of a line segment L is defined according to the distance measure defined between two segments that accounts their *parallel, perpendicular* and *angular* distance, as it was presented in Chap. 6. Following DBSCAN paradigm, if the number of its neighboring segments exceeds the *MinLns* threshold parameter, then all the density-connected segments to L are added to the cluster. Before each density-connected set is considered as a cluster, the *trajectory cardinality* of the cluster, i.e. the number of distinct trajectories from which segments have been derived, is checked whether it exceeds the minimum cardinality threshold.

Finally, at the third step (representative trajectory generation), for each cluster formed the algorithm calculates its representative trajectory. The so-called *Representative Trajectory Generation* (*RTG*) algorithm sweeps the cluster's segments towards their average direction vector, counting the number of line segments that are either the starting or the ending point of a line segment. If the resulted number is equal to or greater than *MinLns* threshold, the algorithm calculates the average coordinate of those points and assigns the average into the set of representative trajectory; otherwise, it proceeds to the next point. To avoid segments that are too close to each other, a smoothing parameter γ is utilized. The outcome of the *RTG* algorithm is a (more or less) smooth linear trajectory that best describes the corresponding cluster.

Algorithm TRACLUS

Input: A dataset of trajectories D={tr$_1$, ..., tr$_N$}

Output: (1) A set of clusters C={C$_1$, ..., C$_O$}, (2) A set of representative trajectories

/* Partitioning phase */

1.   for each tr$_i$ in D do
2.   Execute trajectory partitioning; Get a set S of line segments as the result;
3.   Accumulate S into a set D';

/* Grouping phase */

4.   Execute Line Segment clustering for D'; Get a set C of clusters as the result;
5.   for each C$_j$ in C do
6.   Execute RTG; Get a representative trajectory as the result;

In the case where the objects move under the restrictions of a road network, a different strategy is required as the network distance, the capacity, and the topology of the network drastically affect the potential patterns that may appear. FlowScan and NEAT algorithms, presented below, try to address these issues.

FlowScan. *FlowScan* is one of the first techniques that discover such patterns, called *hot routes*, operating upon trajectories moving on a road network. A hot route can be considered as a general traffic flow pattern of moving objects, which is a sequence of nearby, though not necessarily adjacent, edges that share a high amount of traffic. The hot route discovery algorithm proceeds by first defining all *hot route starts*. An edge *e* is considered as a hot route start, if there are more than (a user-defined threshold) *MinTraffic* objects that start from *r* or when *MinTraffic* or more objects converge at *e* from other edges, however by not accounting traffic from edges that belong to other hot routes. For each hot route start, the related hot routes are extracted following a traffic-density expansion process. The expansion of a hot route is performed only at the last edge (initially, the last is equal to the start). The last edge is grown by attaching the *route traffic density-reachable* edges. An edge *s* is *route traffic density-reachable* from an edge *e*, if there is a chain of *directly density-reachable* edges in between, which also share common traffic for a sliding window of consecutive edges in this chain.

Interestingly, instead of using a spatial distance measure to define the neighborhood of an edge, this technique uses the number of hops (i.e. the minimum number of edges between two edges) to measure closeness in the road network. An edge *s* is *directly density-reachable* from an edge *e*, not only if they are close with respect to the number of hops, but also if they share common traffic.

NEAT. *NEAT* is another road network-aware approach for clustering spatial trajectories of objects moving in a road network. It takes into account the physical constraints of a road network, the network proximity, and the flows between road segments during the clustering procedure, which is accomplished in three phases. In the first phase, called *base cluster formation*, the original trajectories are partitioned into a set of trajectory fragments, called *t-fragments*. A *t*-fragment is considered as a sub-trajectory whose sequential points lie on the same road segment. Then, *base clusters* are formed each one containing the equivalent *t*-fragments that belong to the same road segment. Consider the example shown in Fig. 7.4, where 12 base clusters have been discovered after the completion of the first phase, one for each segment of the road network.

In the *flow cluster formation* phase that follows, base clusters are combined into *flow clusters* (i.e. sequences of base clusters that compose a route in the network) with respect to the *merging selectivity*. Merging selectivity is a weighted function that it takes into account flow, density and road speed factors. Intuitively, the result of this phase consists of three flow clusters, e.g. the ones depicted as F_1, F_2, and F_3 in Fig. 7.4. Finally, at the *flow cluster refinement* phase, a variant of DBSCAN clustering algorithm is employed to compress flow clusters in order to optimize the results. The density-connectivity process is driven by the well-known Hausdorff distance function modified to take into account the network distance. In detail, the distance between two flow clusters is calculated based on the corresponding routes.

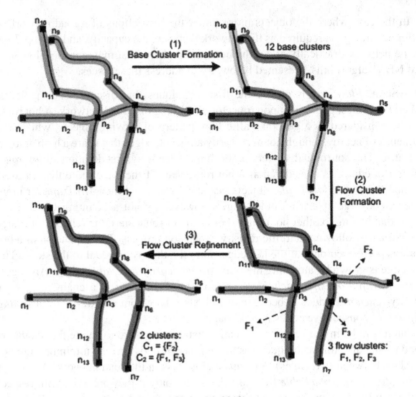

Fig. 7.4 The NEAT framework (*source*: Han et al. 2012)

If the ending locations of two flow clusters do not exceed a predefined distance in terms of network proximity, the clusters are merged into one larger cluster. The final clustering, C_1 and C_2 in Fig. 7.4 (note that F_1 and F_3 are combined into C_2), fulfills high density and continuity criteria, thus depicting frequent traffic flows extracted from mobility data.

7.1.3   Finding Representatives in a Trajectory Dataset

Apart from the clustering process itself, some of the already discussed algorithms provide a side result, which has an added value per se. It is the 'representative' of the cluster, a trajectory (either one of the cluster members or an artificial one) that best represents the members of the cluster. *CenTra* provided by *CenTR-I-FCM* (at its second step) and the *representative trajectory* provided by *TRACLUS* (at its third step) are typical examples.

For example, Fig. 7.5a, b present the *TRACLUS* representatives (the three thick colored lines) and the *CenTR-I-FCM* centroid trajectories (the two differently colored sets of cells), respectively, from a set of synthetic trajectories, representing the

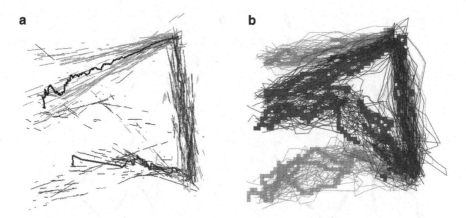

Fig. 7.5 Cluster representatives: (**a**) Representative trajectories produced by TRACLUS; (**b**) Centroid trajectories produced by CenTR-I-FCM

clusters found in the dataset. It is evident that *TRACLUS* discovers three clusters, which are compositions of segments (i.e. sub-trajectories), with a trajectory probably split into different clusters. On the other hand, *CenTR-I-FCM* catches the overall complex mobility patterns.

Both *TRACLUS* representatives and *CenTR-I-FCM* centroids are artificial trajectories for representing collective behavior of trajectories. A different method, called *T-Sampling*, is a deterministic voting methodology that samples the most representative trajectories from a dataset. In other words, the goal of this approach is to effectively sample the top-*k* trajectories that preserve the movement pattern that is hidden inside the whole dataset. More interestingly, the problem is even more challenging when searching for the most representative sub-trajectories, implying a segmentation tactic as a preprocessing phase.

Figure 7.6a illustrates an example of a dataset consisting of four trajectories that have a common lifespan. The segmentation of T_1 according to TRACLUS is illustrated in Fig. 7.6b. On the contrary, according to *T-Sampling*, the segmentation is applied by taking into account the neighborhood of T_1 in the rest of the dataset, yielding a segmentation that is related only on the *representativeness* of the line segments (i.e. reflected by the number of neighboring segments that vote for the line segments of T_1 as the most representatives). In Fig. 7.6c, T_1 is partitioned in three sub-trajectories, where the line thickness is related with the line segments representativeness (i.e. the results of the voting methodology). It holds that the sub-trajectory representativeness values will be almost the same along the line segments of each sub-trajectory. After the partitioning, the goal of sub-trajectory sampling is to select the most representative subset of these sub-trajectories.

By selecting the highest voted segments, which sounds to be an obvious decision, the high-density regions of the dataset will be oversampled, resulting in a non-representative sample, when the aim is to cover the whole space-time extent of the dataset as much as possible. As such, one should take into account the fact that the

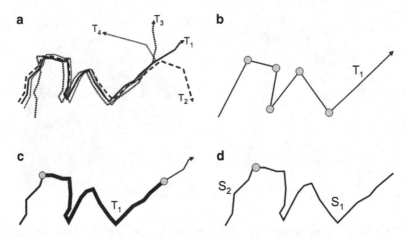

Fig. 7.6 Finding representative sub-trajectories: (**a**) a set of four trajectories; (**b**) partitioning of T_1 using TRACLUS (resulting in six sub-trajectories); (**c**) segmentation of T_1 (three sub-trajectories) using T-Sampling; (**d**) the top-2 representative sub-trajectories produced by T-Sampling

sampling set should contain dissimilar representative sub-trajectories. So, the sampling set should contain high-voted trajectories, which, at the same time, cover the entire three-dimensional space as much as possible. In other words, the resulting set should avoid containing similar sub-trajectories. Although the problem of such a selection process has combinatorial complexity, it has been tackled in *T-Sampling* methodology by a greedy approach, whose intuition is to select a number of trajectories from the initial dataset that their representatives in the sampling set is maximized, according to an optimization function. As such, Fig. 7.6d illustrates the top-2 representative sub-trajectories that best describe the dataset of our running example.

7.2 Moving Clusters for Capturing Collective Mobility Behavior

Unlike the previous clustering techniques that consider time dimension to be continuous, there also exist methods that operate on trajectories, all sampled at the same timepoints (usually with a fixed sampling rate), and only the trajectory samples are considered in the discovery process.

7.2.1 Flocks and Variants

One of the first methods for capturing such collective mobility behavior takes into account the spatial proximity and the direction of the moving objects. For a so-called *flock pattern* to be discovered, a minimal number of trajectories that satisfy such constraints are required. More formally, a *flock* in a time interval I, where I

Fig. 7.7 (a) Flock of four trajectories spanning at five successive timepoints; (b) The lossy-flock problem

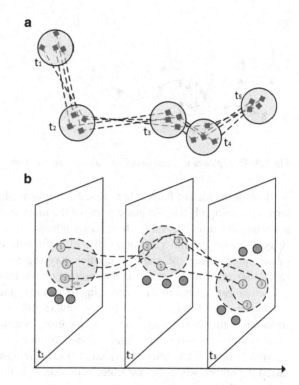

spans for at least k successive timepoints, consists of at least m objects, such that for every timepoint in I, there is a disk of radius r that contains all m entities. Figure 7.7a illustrates an example flock of $m=4$ trajectories in time interval $[t_1, t_5]$.

As expected, parameter r has significant impact on the discovery process of flocks. The selection of a proper disk size turns out to be a difficult issue, as objects that intuitively belong to the flock may be left out due to a slight deviation in their distance from other members of the flock. This is illustrated in Fig. 7.7b, where some grey objects, although outside the chosen disk, could have been part of the flock with respect to prescribed constraints. The main reason for what is referred as the *lossy-flock* problem, is that the sensitive parameter of the disk size is user-defined, instead of being derived from the data distribution. Moreover, if we assume a minimum cardinality of $m=4$, then the outcome is that no flock is found in the dataset. Furthermore, the circular shape is rather restrictive, as ideally, no particular shape should be fixed a-priori.

Setting parameter k of the minimum temporal duration of a flock pattern is also tricky, as a high value might result in identifying very few or even no flock, while a low value might result in over-production of flocks and, more important, not identify long flocks that are of more interest than short ones. Identifying long flock patterns with respect to the time dimension is an interesting extension to the initial problem (called, *top-k longest flock patterns discovery*). Intuitively, the longest flock pattern is the one that has the maximal duration, while the rest of the flock conditions are still satisfied.

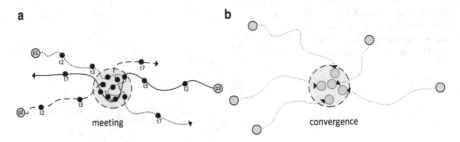

Fig. 7.8 Examples of trajectory patterns: (a) meeting; (b) convergence

Besides the idea of flocks, there have been several attempts to model similar patterns of moving objects. To name a few of the most representative, the discovery of a *meeting* in a time interval I, where I lasts at least k timepoints, consists of at least m objects that stay within a stationary disk of radius r during I. There are two variants of *meetings*: either the same m entities stay together during the entire interval (*fixed-meeting*), or the entities in the meeting region may change during the interval (*varying-meeting*). In Fig. 7.8a, a meeting for p_1 and p_2 and p_3 during $[t_4, t_6]$ is depicted.

On the other hand, the *convergence* pattern, illustrated in Fig. 7.8b, describes trajectories that converge to the same location, coming not necessarily from the same origin. Moreover, the *leadership* pattern characterizes groups of trajectories that move like a flock, and one trajectory (the *leader*) leads the group being located ahead the others, while the rest trajectories compose the set of the *followers* in this variant of the flock pattern.

7.2.2   Moving Clusters

An interesting observation that influences the flock pattern is the identity of the participating objects. If the same m entities stay together during the entire interval, this forms a kind of a *fixed-flock*, while if the entities change during the given interval, a kind of *varying-flock* is formed. Inspired by this idea, the notion of a *moving cluster* was introduced, which is a sequence of clusters $c_1, \ldots, c_k$, such that for each timestamp i, c_i and c_{i+1} share a sufficient number of common objects.

More specifically, assuming a partitioning of the database of the objects' locations with respect to a discretization of the time dimension, a snapshot S_i (i.e. the set of objects and their locations that exist at time t_i), is clustered by using a typical density-based spatial clustering algorithm like DBSCAN, to identify dense groups of objects in S_i, which are close to each other and the density of the group meets the density constraints of the clustering algorithm (e.g. *MinPts* and ε in DBSCAN). Intuitively, if the two spatial clusters at two consecutive snapshots have a large percentage of common objects then they are considered as a moving cluster between these two timestamps. More formally, let c_i and c_{i+1} be two such snapshot clusters for S_i and S_{i+1}, respectively. The sequence $<c_i, c_{i+1}>$ is a moving cluster if

Fig. 7.9 Example of a
moving cluster

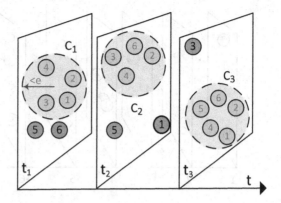

$\dfrac{|c_i \cap c_{i+1}|}{|c_i \cup c_{i+1}|} \ge \theta$, where θ is a percentage threshold for the contents of the two clusters.

Figure 7.9 shows an example of a moving cluster valid during three snapshots. Note that objects may enter or leave the moving cluster during its lifetime.

From the above discussion, it is evident that a moving cluster is defined over a sequence of consecutive timepoints. However, enforcing the consecutive time constraint may result in the loss of interesting patterns. Consider, for example, two moving clusters during $[t_1, t_{i-1}]$ and $[t_{i+1}, t_k]$ containing exactly the same population of objects. The (much longer) moving cluster during $[t_1, t_k]$ was missed because of a (perhaps, very slight) deviation of a few (or even a single) object(s).

7.2.3   Improvements over Flocks and Moving Clusters

On the same line, the idea of density-based spatial clustering, which enables the formulation of arbitrary shapes of groups, has been employed to tackle the lossy-flock problem. In order to bypass the restrictions on the sizes and shapes of the moving clusters and flocks, a number of follow-up techniques have been proposed, including convoys, group patterns, and swarms.

Convoys. A convoy is a group of objects that has at least m objects, which are density-connected with respect to a distance threshold e, during k consecutive timepoints. Convoys are different from moving clusters in that the same objects should be found in consecutive clusters. To demonstrate this, consider clusters C_1, C_2 and C_3 of Fig. 7.10. The three objects found in C_1 form a convoy as they are found together for three consecutive timepoints t_1, t_2, t_3. Instead, if we impose 100 % overlapping clusters, there is no moving cluster, as the common objects between clusters at t_1 and t_2 is only 3/4. For the timepoints t_5, t_6, and t_7, if we impose 50 % overlapping clusters, then C_5, C_6, and C_7 form a moving cluster, but this is not a convoy.

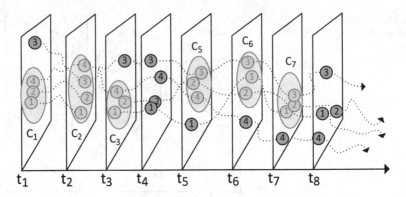

Fig. 7.10 Convoys vs. moving clusters

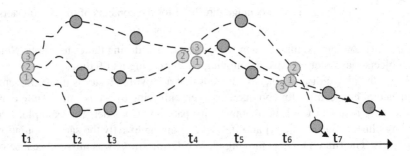

Fig. 7.11 A swarm pattern

All the previously presented approaches require a minimum number of k consecutive timestamps of common behavior (according to various definitions) so as to report a pattern. However, real-world trajectories may meet together at several, though non-consecutive timepoints. For instance, consider the three trajectories illustrated in Fig. 7.11. These trajectories may be considered dissimilar assuming the distance functions discussed in Chap. 6, while they form none of the previously discussed patterns. Nevertheless, when considering timepoints t_1, t_4, and t_6 in our example, the three objects may be considered as moving together. One could think of various nice applications and analyses that would benefit from such kind of movement patterns (e.g. communities meeting together in specific places from time to time). What is needed to meet this challenge is relaxation of the time constraint.

Group patterns. A first approach towards this direction is the so-called *group pattern* that is actually a time-relaxed flock pattern as it consists of a set of moving objects that travel within a radius for certain timestamps that maybe non-consecutive. However, the shortcoming of the group pattern is that it confines the size and shape of the clustered moving objects inside a disk.

Swarms. The above shortcoming has been revised by the concept of swarms. More specifically, a *swarm* is a collection of moving objects with cardinality at least m, that are part of the same cluster for at least k timepoints. Note, that the set of k time-stamps are not required to be consecutive, while the geometric trajectory of each object in-between consecutive or not timepoints, is not imposed under any constraint. According to the above discussion, Fig. 7.11 illustrates a swarm of three trajectories meeting at three non-consecutive timestamps, t_1, t_4, and t_6.

7.3 Sequence Pattern Mining in Mobility Data

Our discussion so far has made clear that trajectories belong to the general class of sequential data. As such, it is a rational step to explore extensions of well-known sequential pattern mining techniques for the case of mobility data. The approaches proposed so far are categorized to two broad classes. The first class studies techniques applicable to trajectories of a single moving object trying to identify regularities in the behavior of the user, while the second class aims at revealing collective sequential behavior of a set of objects.

Periodic patterns. A challenging problem of the first class of methods that has been studied is the extraction of periodic patterns from trajectories, where a *periodic pattern* corresponds to sequences of (in general, non-adjacent) locations that reappear in the trajectory history periodically. In order to be able to apply techniques for mining periodic patterns on sequences of events (or sets of events), each exact location of a trajectory should be considered as a different categorical value. As it is implausible that an object repeats exactly the same sequence of precise locations, a periodic pattern is generalized to be of the form "$r_0 r_1 ... r_{T-1}$", where T is the period of the pattern (that is given as an input parameter) and r_i is a spatial region or the character '*' (implying arbitrary movement in space). For instance, pattern "$HW*H$" is a four-length sequence that appears frequently (i.e. more than a minimum support threshold *min_sup*, according to the frequent pattern mining terminology) in the history of the object, and the repeated sequence that occurs every four timestamps is first at place H (e.g. home), then at place W (e.g. workplace), then at an arbitrary place (hence, '*'), and then at place H again. Interestingly, for a region to be considered as part of a periodic pattern, it should be a dense region, i.e. a region that contains at least a predefined number of trajectory locations. This allows relaxing the spatial granularity shortcoming without the use of a predefined partitioning of the space. An interesting variant of the above-described periodic behavior that has been studied is the discovery of repeating activities at frequently visited regions, but in this case with regular time intervals.

T-Patterns. A different methodology that falls in the second class of mining sequential patterns in mobility data is the so-called *trajectory pattern* (T-pattern), which is defined as a set of trajectories that share the property of visiting the same sequence of places with similar transitions times between each pair of successive places.

Differently from a periodic pattern, a T-pattern is a sequence of visited regions, frequently visited in a specified order with similar transition times. Below, we provide three T-patterns, which are expressed as sequences of pairs, each of which includes the duration required for the trajectories to move from the previous frequently visited region to the current one. For instance, TP_1 implies that 31 trajectories moved from region A (the transition time to the first region is not available) to region B at a typical transition time between 9 and 15 min. One should also note that T-patterns might be overlapping (see TP_1 and TP_3 in the example below).

TP_1: <(), A> <(9,15), B> (supp:31)
TP_2: <(), A> <(4,20), C> (supp:26)
TP_3: <(), A> <(9,12), B> <(10,56), D> (supp:21)

The participating regions in T-patterns could either be defined by the user (implying regions that a domain expert identifies as interesting for the analysis) or be computed automatically by the method itself. More specifically, such *popular regions* are computed following a clustering approach that utilizes the density of the trajectories in the space. The T-pattern algorithm starts by tessellating the space into small rectangular cells, for each of which the density (i.e. the number of trajectories that either cross it or found inside the cell) is computed. Then, by following a region-growing technique, the dense areas are enlarged by including nearby cells as long as the density criterion is fulfilled. Once popular regions have been identified, the algorithm filters the trajectories that intersect these regions and computes the travel time duration between those regions. At this point, the outcome of the processing can feed a sequence-mining algorithm to generate T-patterns followed by a minimal number of trajectories.

Algorithm T-Pattern (with static regions of interest - RoI)

Input: (1) a set of input trajectories T, (2) a grid G_0, (3) a minimum support/density threshold δ, (4) a radius of spatial neighborhoods ε, (5) a temporal threshold τ.

Output: A set of pairs (S, A) of sequences of regions with temporal annotations.

1. G=ComputeDensity(T, G_0, ε);
2. RoI=PopularRegions(G, δ);
3. D=Translate(T, RoI);
4. TAS_mining(D, δ, τ);

We should note that such a required technique is not simply an algorithm for solving the *frequently sequential pattern* problem over a database of sequences (where each element of each sequence is a timestamped itemset), such as PrefixSpan or SPADE, as the sequential patterns should be enriched with the transition times between their elements. This particular kind of sequential pattern mining problem is solved by techniques that are capable of mining *Temporally Annotated Sequences* (TAS).

7.4 Prediction and Classification in Mobility Data

An important data mining task in the mobility data exploration field is that of predicting the future location of trajectories. A naïve solution to the problem that is meaningful in real-time applications is to be based on the current motion vector of the objects
(i.e. their current location along with their velocity vector), and assuming the validity
of this vector for the near future, one may use extrapolation to predict the location of
the moving objects. This approach has been followed by techniques that design access
methods for future queries. Apparently, this approach does not follow any analytical
methodology that is more useful in offline applications where decisions are taken
based on significant evidences found after historical data analysis.

On the other hand, the idea of building a model that recognizes an object and classifies (categorizes or annotates) it with respect to a predefined set of classes (categories or labels, respectively), using a set of training data, has also been applied in
mobility data. This is very useful in applications where the goal of the analysis task is
to predict whether moving objects exhibit some known behavior (e.g. that of a com-
muter or a tourist that visits the city center), in order to advertise appropriate services.
Moreover, as usually done in legacy classification tasks, we could train a two-label
(i.e. normal vs. abnormal) classifier model to detect unusual behavior of moving
objects, i.e. trajectories that are significantly different with respect to others.

7.4.1 Future Location Prediction

Although prediction tasks are usually tackled as special cases of classification techniques, the peculiarity of mobility data has led researchers to use mostly frequent
patterns and association rules, as the step upon which they define the prediction
model. More specifically, one category of techniques extracts association rules from
the trajectories of a single user, which are utilized in a subsequent step to infer user's
next movement. The left-hand side (LHS) as well as the right-hand-side (RHS) of
these association rules actually consist of frequent visited regions by the user. The
best association rule, according to which the prediction is driven, is selected by
using the popular notion of *support*. In detail, the support of a candidate association
rule is measured as a distance measure that quantifies the matching between the
trajectory and a rule. A similar strategy has been followed also by techniques that
use either the whole MOD or the latter appropriately filtered so as to infer some kind
of a global model, upon which the prediction will take place.

WhereNext. The most representative approach of this category is the so-called
WhereNext, and builds on the previously discussed T-patterns, which are used as the
mean to construct the required global model. Precisely, the *WhereNext* methodology
first extracts a set of T-patterns and then constructs a T-pattern tree, which is a
prefix-tree built upon the frequently visited sequence of regions of the discovered
T-patterns. In this process, a T-pattern is considered as a prefix of another T-pattern

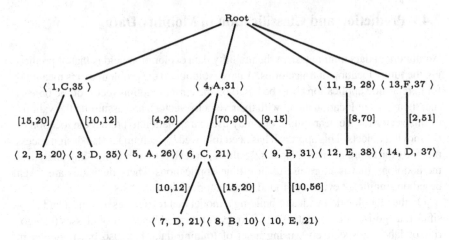

Fig. 7.12 An example of a T-pattern tree (*source*: Monreale et al. 2009)

if and only if the cardinality (i.e. the size/length of the sequence) of the first is equal to or smaller than that of the second and the sequence of all regions of the first is a sub-sequence of that of the second. Interestingly, in order to produce a compact T-pattern tree and in order to account also the transition times and the support values of the original T-patterns, a unification process takes place during tree construction. In this connection, each path of the tree is a valid T-pattern. An example of T-pattern tree is illustrated in Fig. 7.12.

After the T-pattern tree has been built, a given trajectory (for which we want to make a prediction) is used in order to find the best matching score with respect to all paths (i.e. T-patterns) that it fits with. Interestingly, this fitness score takes into account whether the trajectory spatially intersects a node (i.e. a frequent region) of a path of the tree, or, if it does not, it accounts the spatiotemporal distance of the trajectory with respect to the tree node. At the end, the predicted future location of the trajectory is the region that corresponds to the final node of the best path.

7.4.2 Classification and Outlier Detection

A naïve approach to tackle classification tasks in mobility data is to transform trajectories to some feature space and then train a classifier upon it. This way actually under-utilizes the trajectory data, as well as some important sub-trajectories or even the regions wherefrom they pass, which may be more discriminative with respect to the transformed vectors of the resulted feature space.

TRACLASS. Inspired by the above idea, the *TRACLASS* framework proposed a trajectory classification approach that predicts the class label of moving objects based on their trajectories as well as other features. The method is based on the observation that discriminative features might appear as regions at portions of

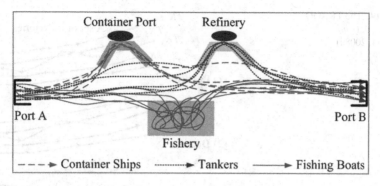

Fig. 7.13   The TRACLASS framework (*source*: Lee et al. 2008b)

trajectories. An example is shown in Fig. 7.13: all trajectories go from Port A to Port B; one group of trajectories passes through a container region; another group passes through the refinery before arriving at Port B; a third group stops at a fishing area before moving to Port B. Taking advantage of this knowledge, we can separate container ships and tankers despite the fact that they move for long in common paths. Moreover, parts of the trajectories moving without a common path in the fishery give as the ability to identify fishing boats. Thus, discriminative features that appear as regions and sub-trajectories cannot be detected in whole trajectories, but they can be extracted by sub-trajectories. Actually, since the method classifies sub-trajectories, it is possible to identify trajectories with different behaviors and different goals.

More precisely, *TRACLASS* first partitions trajectories based on their shapes by using a variant methodology of *TRACLUS* (presented in Sect. 7.1). The difference is that after the discovery of the characteristic points of the trajectories, the method proceeds with a *class-conscious* partitioning. In detail, the rule that drives the partitioning is that trajectory partitions should not exceed the class boundaries. For this purpose, if the most prevalent class around one endpoint is different from that around the other endpoint, then this heterogeneous trajectory partition that occurs among the two endpoints is further partitioned for maintaining the uniformity of the class distribution inside each partition as pure as possible.

The next step in this development is *hierarchical region-based clustering*. The goal is to detect regions containing sub-trajectories mostly of one class within a rectangular region irrespectively of their movement patterns. The discovery of the regions is achieved as an optimal trade-off between *homogeneity* (i.e. purity of the class distribution in each region) and *conciseness* (i.e. minimum number of regions). After all the homogenous regions have been detected, *trajectory-based clustering* is applied to the sub-trajectories of the non-homogeneous regions in order to discover common movement patterns for each of the included classes. Clustering is again based on a variant of TRACLUS with the main difference being that *TRACLASS* incorporates a class constraint that guarantees that a cluster is derived from only one class. Clusters with high discriminative power (so-called, *separation gain*), are selected for effective classification. The separation gain is calculated based on

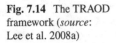

Fig. 7.14 The TRAOD framework (*source*: Lee et al. 2008a)

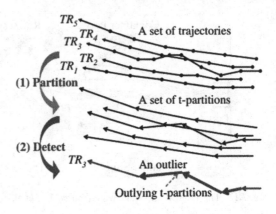

Hausdorff distance that measures the distance from a specific cluster to other clusters of different classes. In order to further improve the accuracy of the classification, clusters are linked together if the ratio of the common trajectories exceeds a threshold. After the above-described process each trajectory is converted to a high-dimensional feature vector, whose i-th entry measures the frequency of occurrence of the i-th feature in the trajectory. A feature in this case is either a region-based cluster or a trajectory-based cluster. Finally, a classifier is build upon the resulted feature space using the support vector machine.

TRAOD. Following the previous line of research, the *TRAOD* framework studies the problem of detecting among a set of trajectories, those that behave differently from their neighbors (hence, considered outliers). The motivation again is based on the fact that due to the complex path that trajectories may follow, comparing entire trajectories cannot lead to the identification of outlying portions. As shown in Fig. 7.14, a sub-trajectory of TR_3 behaves differently from its neighbors, and therefore may be considered as outlier. *TRAOD* discovers outliers by following a partitioning and, then, a detection phase. Initially, the partitioning phase is applied by following a similar approach with the one used in *TRACLUS* and *TRACLASS*. A so-called *t-partition* is considered as outlier if there are not sufficient similar neighboring trajectories with respect to the *TRACLUS* distance function (presented in Chap. 6). In turn, a trajectory is close to a t-partition if a sufficient portion of a trajectory is close to it. A trajectory becomes an outlier if it contains sufficient number of outlying t-partitions (i.e. not enough similar neighbors). In dense regions, t-partition has a lot of close trajectories, thus preventing the detection of the corresponding outlying ones. An appropriate coefficient weights the impact of density in both dense and sparse regions. Due to the flexibility of the distance function that uses different features, this approach is capable of detecting outliers with respect to the location (i.e. positional outliers) as well as the direction of the trajectories (i.e. angular outliers).

ROAM. Sometimes, anomalies can be identified in various levels of abstraction that correspond to diverse spatial and temporal granularities. *ROAM* (Rule and

Motif-based Anomaly Detection in Moving Objects), is an automated approach for detecting peculiar behavior in movements with such characteristics. *ROAM* methodology suggests the transformation of the trajectories to a *motif-based feature space*. Trajectories are partitioned into fragments, which when overlaid, may lead to the discovery of common patterns, called *motifs*. Movement motifs resemble geometric transformations, such as straight lines, U-turns, loops, etc. Motifs are further enriched with other spatiotemporal attributes, such as starting and ending time, location, speed, etc., constituting *motif expressions*, which are used for setting the feature space. As such, given a set of motifs and a trajectory, the latter can be transformed (by traversing each trajectory following a sliding-window approach) to a sequence of motif expressions. Here is an example of a motif-trajectory as a sequence of *<motif, attribute, attribute value>* triplets:

<straight, time, 8am>, <U-turn, time, 9pm>, ...

..., <U-turn, speed, 5kmph>, <loop, time, 10am>

After motif expressions have been determined, motif-trajectories are converted into vectors in the feature space. Actually, the i-th feature of the vector corresponds to the number of occurrences motif x's attribute j expressed a value k in the trajectory. In order to decrease the dimensionality of the feature space, *automated hierarchy extraction* is applied that generalizes motif features into higher levels of abstraction. For example, features <left-turn, time, 10pm> and <left-turn, time, 11am> can be generalized into <left-turn, time, morning>. Motif attribute values are discretized in case of numerical attributes or clustered to form a multi-resolution hierarchic view of the data. For spatial attributes, this is achieved by performing hierarchical micro-clustering where the centroid of each micro-cluster corresponds to a new feature. These extracted features substitute the original features, thus forming a new feature space. At the final step, a *hierarchical-based classifier* is trained by exploring the aforementioned hierarchies.

7.5 Summary

Knowledge discovery in trajectory databases discovers behavioral patterns of moving objects that can be exploited in several fields. In this chapter, we provided insights to a representative set of methods that have been proposed in the already rich literature of the mobility data mining domain. The data mining models discussed in this chapter have found nice applications in active domains of knowledge discovery.

Although many interesting mining models are out there to represent various types of collective behavior of moving objects, we feel that the field has still many steps to follow. Some interesting directions that we foresee is the tighter integration of the ad-hoc (so far) mining methods with MOD engines (e.g. register an efficient trajectory clustering method as a query operator in the MOD), not only in order to take advantage of the nice features provided by DBMS (indexing, query processing and optimization, etc.), but also to facilitate explorative analysis, which is a key issue in mobility data mining. Towards this direction, the field of geospatial visual analytics has already shown nice synergies, as without effective visualization of the mobility data, the analysis of such kind of data becomes cumbersome. Another hot topic for researchers that are challenged by the idea of mining mobility data, but also for practitioners of the field, is the bridging between semantic aspects of the mobility data, but also the meeting with the Big Data era. Both of these two important issues are discussed in detail in subsequent chapters, namely Chaps. 9 and 10, respectively.

7.6 Exercises

Ex. 7.1. Consider a trajectory distance function that only takes into account the starting positions of the trajectories and ignores the rest of the positions. Discuss the application of such a naïve distance function into T-OPTICS clustering algorithm instead of the sophisticated one that T-OPTICS uses. How would the clustering result of Fig. 7.2a change?

Ex. 7.2. Discuss the effect of applying the TRACLUS sub-trajectory clustering algorithm to the toy dataset of Fig. 7.6a.

Ex. 7.3. One way to analyze a MOD in an explorative way is to store the outcome of your intermediate analysis (actually, the output of some mobility mining algorithms), give this as input to a subsequent step, and so on. How would you store patterns like flocks, moving clusters, and T-patterns in a legacy DBMS? How would you index these spatiotemporal patterns and what kind of queries/analyses would you be able to apply? Following the approach of MOD, think of new datatypes that could be used to store such mobility patterns in a compact way.

Ex. 7.4. Explain the differences between moving clusters and convoys. How would you adapt the methodology of moving cluster discovery (outlined in Sect. 7.2.2) in order to identify convoys (outlined in Sect. 7.2.3)? Think of other feasible transformations between pairs of patterns discussed in Sect. 7.2.

Ex. 7.5. Devise an algorithm that takes as input a trajectory dataset and a set of T-patterns and returns those trajectories from the dataset that conform to these T-patterns. Apply this algorithm in a trajectory dataset of your choice.

Ex. 7.6. Consider the three example T-patterns presented in Sect. 7.3. How the T-pattern tree of the WhereNext prediction methodology would look like, when constructed by using only these patterns?

Ex. 7.7. Think of a rich palette of motifs that would be of special interest in the application area of vessels' movements. Devise an algorithm that extracts these motifs for each vessel's trajectory. Apply this algorithm on a real vessels trajectory dataset, such as the ones found in ChoroChronos.org dataset repository.

7.7 Bibliographical Notes

The literature on mobility data mining, overviewed earlier in Giannotti and Pedreschi (2008), can be classified in various fields depending on the mining models used to discover the indeed rich collection of patterns described earlier. As such, there have been proposed approaches that identify various types of clusters of moving objects. Gaffney and Smyth (1999) and Cadez et al. (2000) proposed probabilistic algorithms that cluster small trajectories, Nanni and Pedreschi (2006) extended density-based clustering techniques and introduced the T-OPTICS algorithm presented in Sect. 7.1.1, while Pelekis et al. (2009, 2011) proposed CenTR-I-FCM for clustering uncertain trajectories, also presented in Sect. 7.1.1. Contrary to the above algorithms, which consider entire trajectories, the TRACLUS framework, presented in Sect. 7.1.2, was the first sub-trajectory clustering algorithm (Lee et al. 2007). TRACLUS also proposed an algorithm for computing a virtual representative trajectory for each cluster.

In case where moving objects move under the restrictions of a transportation network, FlowScan and NEAT approaches presented in Sect. 7.1.2, were proposed by Li et al. (2007) and Han et al. (2012), respectively. In the same field, Sacharidis et al. (2008) proposed an online approach to discover and maintain hot motions paths while Chen et al. (2011) tackled the problem of discovering the most popular route between two locations based on the historical behavior of travelers.

The T-Sampling technique for selecting the representative trajectories or sub-trajectories from a dataset via sophisticated sampling methodologies, outlined in Sect. 7.1.3, was proposed in Pelekis et al. (2010) and Panagiotakis et al. (2012), respectively.

In Laube and Imfeld (2002), Laube et al. (2004, 2005), the authors define various mobility behaviors around the idea of the flock pattern, such as the leadership, convergence and encounter patterns. Gudmundsson and van Kreveld (2006), Gudmundsson et al. (2007) extended previous works in order to find long flock patterns, including the time dimension. Moving clusters presented in Sect. 7.2.2 were introduced in Kalnis et al. (2005), where various algorithms were proposed for their efficient computation. A similar line of research has been followed by Li et al. (2004) who studied the notion of *moving micro cluster*, which is a group of objects that are not only close to each other at current time, but they are also expected to move together in the near future. Spiliopoulou et al. (2006) studied transitions in moving clusters (e.g. disappearance and splitting) between consecutive time points, while Jensen et al. (2007) proposed techniques for maintaining clusters of moving objects by considering the clusters of the current and near-future positions.

Among the models discussed in Sect. 7.2.3, the convoy pattern and methods for its efficient computation were studied in Jeung et al. (2008b) while Wang et al. (2006) defined the group pattern, a generalization of which was proposed in Li et al. (2010a), where the swarm pattern was introduced.

The most representative approaches that tackle sequential trajectory pattern mining problems have been presented in Cao et al. (2005, 2006, 2007), Giannotti et al. (2007), and Li et al. (2010b). As already discussed, researchers challenged by the future location prediction problem, based on such sequential patterns to tackle this problem. Prediction techniques that utilize association rules extracted from the trajectories of a single user are proposed in Yavas et al. (2005) and Jeung et al. (2008a), while others that base the prediction by using the movements of all the moving objects in a database are presented in Morzy (2007) and Monreale et al. (2009).

One of the few works that study classification in mobility data is the TRACLASS framework (Lee et al. 2008b). An approach that shares ideas with TRACLASS is the TRAOD framework, which proposes a sub-trajectory outlier detection algorithm (Lee et al. 2008a). The ROAM rule- and motif-based anomaly detection algorithm has been proposed by Li et al. (2007). These techniques were presented in Sect. 7.4.2.

Finally, there are few recent works that extract patterns at a higher level of abstraction, by considering the semantics of trajectories (Bogorny et al. 2009, 2011) and by defining data mining query languages for mobility data (Trasarti et al. 2011; Giannotti et al. 2011).

References

Bogorny V, Kuijpers B, Alvares LO (2009) ST-DMQL: a semantic trajectory data mining query language. Int J Geogr Inf Sci 23(10):1245–1276

Bogorny V, Avancini H, De Paula BL, Kuplish CR, Alvares LO (2011) Weka-STPM: a software architecture and prototype for semantic trajectory data mining. Trans GIS 15(2):227–248

Cadez V, Gaffney S, Smyth P (2000) A general probabilistic framework for clustering individuals and objects. In: Proceedings of SIGKDD

Cao H, Mamoulis N, Cheung DW (2005) Mining frequent spatio-temporal sequential patterns. In: Proceedings of ICDM

Cao H, Mamoulis N, Cheung DW (2006) Discovery of collocation episodes in spatiotemporal data. In: Proceedings of ICDM

Cao H, Mamoulis N, Cheung DW (2007) Discovery of periodic patterns in spatiotemporal sequences. IEEE Trans Knowl Data Eng 19(4):453–467

Chen Z, Shen HT, Zhou X (2011) Discovering popular routes from trajectories. In: Proceedings of ICDE

Gaffney S, Smyth P (1999) Trajectory clustering with mixtures of regression models. In: Proceedings of KDD

Giannotti F, Pedreschi D (2008) Mobility, data mining and privacy, geographic knowledge discovery. Springer, Berlin

Giannotti F, Nanni M, Pinelli F, Pedreschi D (2007) Trajectory pattern mining. In: Proceedings of KDD

Giannotti F, Nanni M, Pedreschi D, Pinelli F, Renso C, Rinzivillo S, Trasarti R (2011) Unveiling the complexity of human mobility by querying and mining massive trajectory data. VLDB J 20(5):695–719

Gudmundsson J, van Kreveld MJ (2006) Computing longest duration flocks in trajectory data. In: Proceedings of GIS

Gudmundsson J, van Kreveld MJ, Speckmann B (2007) Efficient detection of patterns in 2d trajectories of moving points. GeoInformatica 11(2):195–215

Han B, Liu L, Omiecinski E (2012) NEAT: road network aware trajectory clustering. In: Proceedings of ICDCS

Jensen CS, Lin D, Ooi BC (2007) Continuous clustering of moving objects. IEEE Trans Knowl Data Eng 19(9):1161–1174

Jeung H, Liu Q, Shen HT, Zhou X (2008a) A hybrid prediction model for moving objects. In: Proceedings of ICDE

Jeung H, Yiu ML, Zhou X, Jensen CS, Shen HT (2008b) Discovery of convoys in trajectory databases. In: Proceedings of VLDB

Kalnis P, Mamoulis N, Bakiras S (2005) On discovering moving clusters in spatio-temporal data. In: Proceedings of SSTD

Laube P, Imfeld S (2002) Analyzing relative motion within groups of trackable moving point objects. In: Proceedings of GIScience

Laube P, van Kreveld M, Imfeld S (2004) Finding REMO—detecting relative motion patterns in geospatial lifelines. In: Proceedings of SDH

Laube P, Imfeld S, Weibel R (2005) Discovering relative motion patterns in groups of moving point objects. Int J Geogr Inf Sci 19(6):639–668

Lee JG, Han J, Whang KY (2007) Trajectory clustering: a partition-and-group framework. In: Proceedings of SIGMOD

Lee JG, Han J, Li X (2008a) Trajectory outlier detection: a partition-and-detect framework. In: Proceedings of ICDE

Lee JG, Han J, Li X, Gonzalez H (2008b) TraClass: trajectory classification using hierarchical region-based and trajectory-based clustering. Proceedings VLDB 1(1):1081–1094

Li Y, Han J, Yang J (2004) Clustering moving objects. In: Proceedings of KDD

Li X, Han J, Lee JG, Gonzalez H (2007) Traffic density-based discovery of hot routes in road networks. In: Proceedings of SSTD

Li Z, Ding B, Han J, Kays R (2010a) Swarm: mining relaxed temporal moving object clusters. Proceedings VLDB 3(1):723–734

Li Z, Ding B, Han J, Kays R, Nye P (2010b) Mining periodic behaviors for moving objects. In: Proceedings of KDD

Monreale A, Pinelli F, Trasarti R, Giannotti F (2009) WhereNext: a location predictor on trajectory pattern mining. In: Proceedings of KDD

Morzy M (2007) Mining frequent trajectories of moving objects for location prediction. In: Proceedings of MLDM

Nanni M, Pedreschi D (2006) Time-focused clustering of trajectories of moving objects. J Intell Inf Syst 27(3):267–289

Panagiotakis C, Pelekis N, Kopanakis I, Ramasso E, Theodoridis Y (2012) Segmentation and sampling of moving object trajectories based on representativeness. IEEE Trans Knowl Data Eng 24(7):1328–1343

Pelekis N, Kopanakis I, Kotsifakos E, Frentzos E, Theodoridis Y (2009) Clustering trajectories of moving objects in an uncertain world. In: Proceedings of ICDM

Pelekis N, Kopanakis I, Panagiotakis C, Theodoridis Y (2010) Unsupervised trajectory sampling. In: Proceedings of ECML-PKDD

Pelekis N, Kopanakis I, Kotsifakos E, Frentzos E, Theodoridis Y (2011) Clustering uncertain trajectories. Knowl Inf Syst 28(1):117–147

Sacharidis D, Patroumpas K, Terrovitis M, Kantere V, Potamias M, Mouratidis K, Sellis T (2008) On-line discovery of hot motion paths. In: Proceedings of EDBT

Spiliopoulou M, Ntoutsi I, Theodoridis Y, Schult R (2006) MONIC: modeling and monitoring cluster transitions. In: Proceedings of KDD

Trasarti R, Giannotti F, Nanni M, Pedreschi D, Renso C (2011) A query language for mobility data mining. Int J Data Warehousing Mining 7(1):24–45

Wang Y, Lim EP, Hwang SY (2006) Efficient mining of group patterns from user movement data. Data Knowl Eng 57(3):240–282

Yavas G, Katsaros D, Ulusoy O, Manolopoulos Y (2005) A data mining approach for location prediction in mobile environments. Data Knowl Eng 54(2):121–146

Chapter 8
Privacy-Aware Mobility Data Exploration

The increasing availability of data due to the explosion of mobile devices and positioning technologies has led to the development of efficient mobility data management and mining techniques. However, the analysis of such data may enhance significant risks regarding individuals' privacy. Consider for example a user requesting a service for nearby points of interest (POI), such as restaurants or pharmacies. Even if hiding user identifier, the request contains enough information to identify the requester. By linking exact coordinates sent to the service provider with public available information about POI's, a third party can increase the probability that the request was sent e.g. from user's home. Consequently, location data should be kept confidential since its disclosure may represent a brutal violation of privacy protection rights. Moreover, developing techniques able to analyze and extract significant patterns from traces left by moving objects can provide insight to the data holders and support to decision-making and strategic planning activities (consider, for instance, patterns depicting typical movement behavior of people moving in an urban environment and how these patterns evolve over time). For this reason, publishing mobility data for analysis purposes is an unavoidable need. But what kinds of privacy threats rise if a MOD is released? By linking an anonymous MOD with public available information, is a malevolent user able to conclude personal behaviors or, even worse, uniquely re-identify the user behind a trajectory? This chapter provides a survey regarding privacy-preservation techniques for location and moving object data. In particular, we discuss the challenges with respect to privacy on mobility data, focusing on three categories of privacy-preservation techniques, namely (a) privacy in the context of *Location-based Services* (*LBS*), where a trusted server aims at providing the service without threatening the anonymity of the user requiring the service, (b) *privacy-preserving mobility data publishing*, where the goal is to release a sanitized version of the original MOD for public use, and (c) *privacy-aware mobility data querying*, where the focus is on providing anonymous answers to queries posed by the users to a MOD that is maintained in-house.

N. Pelekis and Y. Theodoridis, *Mobility Data Management and Exploration*,
DOI 10.1007/978-1-4939-0392-4_8, © Springer Science+Business Media New York 2014

8.1   Privacy in Location-Based Services

An individual's location data contains her identity along with her time-stamped positions. Given a set of individuals, the concern of location privacy arises when location information about a specific individual is disclosed. Location privacy is a particular type of privacy and is defined as *the ability to prevent untrusted third parties to reveal current or past location of an individual*. Location data, due to its peculiar nature, enables delicate privacy issues, since intrusive inferences about sensitive personal information might be enabled.

An approach to maintain users' anonymity is preventing de-identification by using *perturbation* and *obfuscation* techniques; in this case, users are protected since the reported location is represented with a fake value (the accuracy and the amount of privacy mainly depend on how far is the reported location from the exact location). Another approach, called *cloaking*, follows the generalization paradigm from relational data and generalizes space (and, eventually, time) until the spatial (or spatiotemporal) region produced contains at least K users; in this case, the user's location is represented as a region (that includes the actual user location) and the probability that a malevolent is able to achieve user identification does not exceed $1/K$.

The communication of a user's position to a LBS provider upon a service request may enhance privacy threats based on the assumption that service providers are untrusted. Thus, a trusted server (anonymizer) is responsible for hiding sensitive user location information. As shown in Fig. 8.1, when a user poses a query, the trusted server first removes the obvious identifiers and then blurs the actual location of the user into an area containing at least K users, called *Anonymized Spatial Region* (ASR). It is the ASR that is transmitted to the LBS provider, which in turn returns a candidate result set satisfying the query condition for any possible point within this area. Then, the anonymizer refines the candidate results to

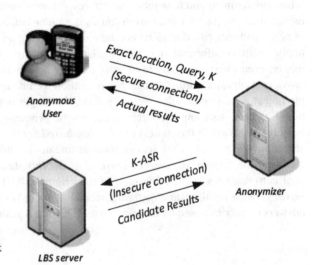

Fig. 8.1 Privacy framework in LBS

provide the actual ones (recall that the anonymizer is aware of the requester's exact location, which means that it is trusted).

Consider, for example, that a requester submits an LBS request from her device asking for the nearest hospital. She does not want to reveal to an eventual adversary her intention. Through a trusted server that removes user identification, the query is transmitted to the LBS server. However, the latter needs the requester's exact coordinates in order to retrieve the nearest hospital, so this location should somehow be communicated. On the other hand, taking into consideration that the LBS server is not trusted, an adversary might be able to follow the communication and disclose the requester's identity. An idea to overcome this problem is that the anonymizer (which is aware of the current location of all LBS users) identifies $K-1$ other users in the requester's neighborhood, encloses all K locations into a K-ASR, and forwards this region to the LBS server as the 'location' of the requester. In turn, the LBS server identifies all closest hospitals to 'any' location within this region and sends back the candidate answer. Finally, the anonymizer is able to filter this set and provide the correct answer to the requester.

In general, there are two LBS categories. The first category corresponds to LBS offering a service in a single communication with the user, known as *snapshot queries*. On the other hand, there are cases where users request for a service along their movement; this is known as *continuous queries*. In the rest of this section, various privacy-preserving approaches, either snapshot or continuous, will be discussed.

8.1.1 Privacy in Snapshot LBS

The most popular anonymization methodology for the protection of individuals' requests in LBS is *cloaking*. A location anonymizer is responsible to gather users' locations and to blur them into K-anonymous cloaked spatial regions with respect to user specified anonymization level and minimum spatial region area. All the algorithms belonging to this class aim at satisfying K-anonymity principle, by generalizing the location of requester. This is achieved by forming an anonymized area (ASR), which covers the requester plus other $K-1$ users.

An example of cloaking is illustrated in Fig. 8.2a. U_1 is asking for the nearest hospital. The anonymizer identifies two other users in U_1's neighborhood (U_2 and U_3), encloses the three locations into a 3-ASR, and forwards this region to the LBS server as the 'location' of the requester. In turn, the LBS server sends back the candidate answer $\{H_1, H_2, H_3, H_4\}$, out of which, the anonymizer is able to provide the correct answer (H_2) to U_1.

As exemplified in Fig. 8.2a, the $K-1$ users selected to form the ASR are the $K-1$ nearest neighbors to U_1; technically, this approach is called *Center Cloak*. However, this approach suffers from the so-called *center-of-cloaked-area attack*, since the user to lies closest to the center of ASR is the most probable to be the query requester. To overcome this shortcoming, a variant called *NN-Cloak* finds an ASR containing the user plus another K users, as follows: it randomly chooses one of the requester's

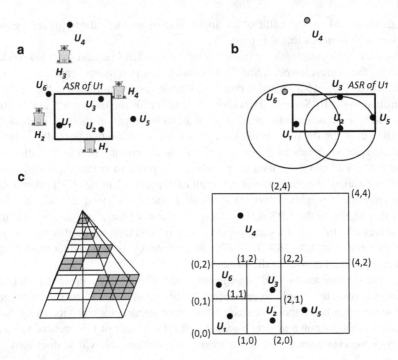

Fig. 8.2 Examples of spatial cloaking for snapshot LBS: (**a**) *Center Cloak*; (**b**) *NN-Cloak*; (**c**) grid partitioning (*Casper*)

$K-1$ nearest neighbors and, in turn, retrieves her $K-1$ nearest neighbors. So the ASR is formed as the MBR containing the locations of the requester and of those K users. This variant is illustrated in Fig. 8.2b: first, the two nearest neighbors to U_1 (U_2 and U_6) are retrieved; second, U_2 is randomly chosen between the two; third, the two nearest neighbors to U_2 (U_3 and U_5) are retrieved; fourth, the 4-ASR containing $\{U_1, U_2, U_3, U_5\}$ is formed and submitted to the LBS server.

On the other hand, grid-based methodologies create the ASR by partitioning the overall space covered by the trusted server in a grid fashion, in such a way that each cell contains at least K users. *Casper* is a well-known technique of this category, based on Quad-trees. Users' locations are indexed in a Quad-tree fashion that decomposes space into various structure layers as shown in Fig. 8.2c. When a user poses a query, *Casper* retrieves the corresponding lowest-level cell. If K users are contained in this cell, then the K-ASR corresponds to that grid cell; otherwise, augmentation is performed by including the sibling (horizontal and vertical) cells or, if needed, by working at the parent level, until a K-ASR is found. In Fig. 8.2c, U_1 belonging to quadrant <(0,0), (1,1)> poses a query where $K=3$. *Casper* detects that the specific cell contains only one user, so it examines sibling cells and defines <(0,0), (2,2)> to be the 3-ASR since it contains at least three users, in our example $\{U_1, U_2, U_3, U_6\}$.

Fig. 8.3 An example of *Probabilistic Cloak* (**a**) exact user locations and (**b**) cloaked user locations

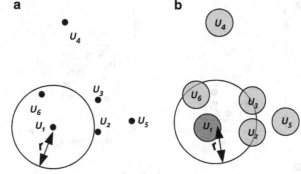

In contrast to previous techniques that are based on the generalization of LBS requester's location (using ASR), spatial obfuscation approaches are based on location perturbation. As a representative of this class, *Probabilistic Cloak* introduces imprecision exploiting on users' privacy profiles preferences (for instance, a user may demand to be located with inaccuracy of at most 200 m—i.e. in a circle with radius 200 m—whereas another user may enforce zero inaccuracy relaxing, of course, privacy concerns in her case). As such, the ASR of the requester is computed irrespectively of the number of the users that are located in her neighborhood, while query location is uniformly distributed into uncertainty area. In the example illustrated in Fig. 8.3, user U_1 wants to identify if there exist other users within distance r. The cloaked regions of each neighbor are calculated based on the corresponding privacy preferences. The algorithm returns as an answer to the queries, the candidate result set $\{U_2, U_3, U_6\}$ along with the probabilities calculated as the percentage of the overlap between the region of the request and each cloaked region.

8.1.2   Privacy in Continuous LBS

Introducing mobility of users makes the privacy-preserving issue even more demanding. Upon a user's request, a group-based cloaking approach formulates groups of $K-1$ nearby users and defines ASR accordingly. The overhead of this approach is that the members of the group are not 'allowed' to leave it until the requested service is completed. To achieve it, as soon as the requester updates her position the same is required by the other members of the group in order for ASR to be up-to-date.

An example of *group-based approach* for continuous LBS is illustrated in Fig. 8.4a. U_1 requests for a continuous service at time t_1. Assuming $K=3$, a group containing $\{U_1, U_3, U_4\}$ is formed at t_1 and the 3-ASR is their locations' enclosing MBB. Moreover, as soon as U_1 updates her location (e.g. at t_2 and t_3), ASR is reformulated, also according to the other two users' locations. Clearly, this approach has the drawback that all users have to transmit their location as soon as it changes, even though they do not request for a service. Moreover, ASR might become very wide (consider the case where the members of the group get far from each other as time passes).

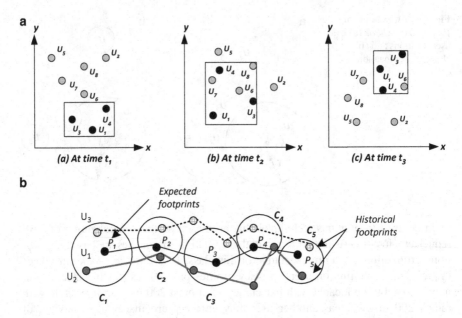

Fig. 8.4 Examples of spatial cloaking for continuous LBS: (**a**) group-based cloaking, (**b**) prediction-based cloaking

As an alternative, a *prediction-based approach* is based on 'footprints', i.e. historical location samples. The entire space is partitioned into equally-sized cells, each one containing K users that have visited this cell sometime in the past. Given a set of footprints that the requester is predicted to follow, the trusted server cloaks them by retrieving the historical footprints of $K-1$ nearby users. Consider the example in Fig. 8.4b with the requester's (U_1) predicted footprints. Assuming $K=3$, the trusted server retrieves historical footprints from nearby users U_2 and U_3. For each predicted footprint p_i of U_1, the region that contains this footprint along with two other footprints from U_2 and U_3 is calculated. The result is a sequence of spatial cloaked regions.

In contrast to the above generalization techniques, the main goal of *data perturbation* techniques is to confuse the exact location of a user that requests a continuous LBS. Let us consider that the entire space is partitioned into *application zones* and *mix-zones*, defined as follows: as soon a users enter a mix-zone, their identification changes to a new, unused pseudonym. In addition, during their stay there, their location information is blocked from being sent to any location-based application. As such, when an adversary finds a user exiting the mix-zone, she cannot distinguish that specific user from any other user who was also located in the mix-zone at that time. The main result is that the adversary is unable to link the identities of users entering with those exiting the mix-zone. Consider the example illustrated in Fig. 8.5: U_1, U_2 and U_3 are the identifiers of three users entering the mix-zone whereas U_x, U_y and U_z correspond to users exiting the mix-zone. Who is who (e.g. $x=1$, and so on) cannot be inferred due to the perturbation resulted in the mix-zone.

Fig. 8.5 Application vs.
mix-zones aiming at data
perturbation

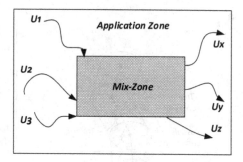

8.2 Privacy Preserving Mobility Data Publishing

Consider a MOD consisting of trajectories of users moving in an urban environment
for a long period of time. By simply removing user identifiers, individuals' privacy
is still under threat. If, for instance, an adversary observes that repeatedly a user X
starts every morning around 8 AM from location A, moves to location B, stays there
for about 8 h, and then returns back to point A, then she can safely assume that
points A and B correspond to the home place and workplace, respectively, of user
X, which may be sensitive personal information. Thus, there appears an essential
need to develop techniques aiming at providing privacy-preservation in MOD.
Privacy-preserving mobility data publishing techniques aim at publishing a sanitized version of the original MOD while maintaining, as much as possible, data
quality. They are mostly based on the K-anonymity principle, according to which a
MOD is anonymized when every trajectory in the dataset is indistinguishable from
at least $K-1$ other trajectories with respect to the values of the quasi-identifiers.
(Note that the K-anonymity principle has been already discussed in Chap. 2 as background knowledge from the relation domain).

Due to the complicated nature of mobility data, the discovery of quasi-identifiers
(which are necessary to be defined in order for the mechanism of K-anonymity to be
performed) is a difficult task, in contrast to relational data. The dependence between
consecutive points of a user's trajectory, both in spatial and temporal dimension,
does not allow to determine a specific set of locations and the corresponding time
intervals to be the quasi-identifiers. Moreover, sensitive locations may vary between
users; a given location (e.g. a hospital) may be meaningless for one user but highly
sensitive for another. Since the definition of quasi-identifiers is not obvious, some
anonymization techniques consider each point of trajectories as sensitive and try to
anonymize trajectories as a whole. An adversary might be able to uniquely identify
a user in every location of her path at any time interval. However, several other
approaches try to deal with various assumptions depicting malevolent knowledge as
will be further explained.

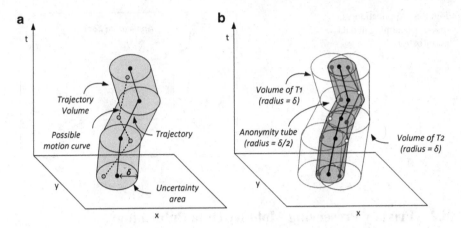

Fig. 8.6 The motivation of NWA technique: (**a**) a cylinder of radius δ anonymizes the trajectory that lies inside the, (**b**) a $(2,\delta)$-anonymous set consisting of two co-localized trajectories and represented as a cylinder of radius $\delta/2$

Following the main research lines of relational anonymization, privacy-aware data publishing techniques include *generalization-* and *perturbation-based* approaches; in the former, the spatio-temporal information is generalized in order to confuse the adversary while in the latter this is achieved by altering (perturbing) the respective information. Algorithms *Never Walk Alone* and *Always Walk with Others*, to be presented in detail in the following sections, are typical techniques that borrow ideas from generalization and perturbation.

8.2.1 Never-Walk-Alone (NWA)

Assume that instead we publish the actual trajectory of a moving object, we cluster together trajectories and for every cluster we publish a cylinder-shaped 'trajectory' of radius δ where all actual trajectories of the cluster lie inside. What we can infer about the moving object that belongs to this cluster is that it moves inside the cylinder but we do not know exactly where; moreover, if at least K trajectories lie inside the cluster, then we have achieved (K,δ)-anonymity. The main assumption is that two trajectories are indistinguishable from each other if they move inside the same cylinder (*uncertainty area*). So the goal is to cluster together groups of at least K co-localized trajectories with respect to a radius δ to form a (K,δ)-anonymized aggregative trajectory. This is graphically illustrated in Fig. 8.6 and it is the motivation behind the *Never Walk Alone* (NWA) technique.

Formally, two trajectories T_1 and T_2, which are both alive during $[t_1, t_n]$ are co-localized with respect to distance δ, if the Euclidean distance between their corresponding points at the same timestamps is less than or equal to δ. A set S of trajectories is (K,δ)-*anonymous* if there exist some so-called *pivot trajectories* T_p, so that at least K trajectories in S are co-localized with the same T_p with respect to an uncertainty radius $\delta/2$. In the example of Fig. 8.6b, the trajectories T_1 and T_2 form a $(2,\delta)$-anonymous set since (a) they are co-localized with respect to distance δ and (b) both are co-localized with a pivot T_p with respect to distance $\delta/2$.

More specifically, the workflow of *NWA* consists of three steps:

1. *Pre-processing step*: the set S of trajectories is partitioned into equivalence classes, with each containing trajectories alive during the same time interval (i.e. having the same starting and ending times). Since such partitioning may not be efficient and in order to increase the cardinality of each class, trajectories are pruned, given a parameter π, which ensures that only one timestamp every π time units can be considered as starting or ending point of a trajectory.
2. *Clustering step*: within each equivalence class found at the previous step, a greedy clustering algorithm is applied, whose objective is to discover the best pivot trajectories around which clusters can be formulated. Every trajectory in the equivalence class is evaluated as pivot in order to form clusters of K trajectories around it, by gathering $K-1$ nearest neighbors that have to obey some proximity constraints (i.e. each cluster should have a radius up to a *max_radius* threshold). The main goal of this step is to keep the radius of produced clusters small, at the price of suppressing some outlier trajectories.
3. *Spatial translation step*: each cluster found in the previous step is transformed into a (K,δ)-anonymous set by applying the minimum necessary space translation on the trajectories of the cluster, so that after translation all trajectories lie within the same uncertainty cylinder of radius $\delta/2$ (in other words, trajectory points not lying inside the cylinder are space translated to the cylinder boundary). Thus, the anonymized set S' of trajectories is produced and released.

8.2.2 Always-Walk-with-Others (AWO)

Let us assume that an adversary is aware of either the part of a trajectory and tries to reveal the entire trajectory or the entire trajectory and is interested in revealing sensitive information about the user behind this trajectory. It is clear that users' privacy is threatened. A solution to this problem is the perturbation of trajectory information. To this line, a perturbation-based anonymization technique, called *Always Walk with Others (AWO)* has been proposed. *AWO* first partitions the original set of trajectories into clusters of K trajectories each, with each cluster being represented as a set of sequences of *K-anonymized regions*. Then, exploiting on these regions, for each cluster it reconstructs K trajectories, thus releasing in the end an anonymized set of reconstructed trajectories that can be considered similar (but by no means identical) to the original. In fact, an atomic dataset is recreated from the anonymized dataset by uniformly selecting atomic points from anonymized regions.

In detail, *AWO* consists of two main steps:

1. *Anonymization step*. Given a set S of trajectories, on each iteration of the step, the algorithm creates an (initially) empty group G, randomly selects a trajectory T_i from S, inserts it into G, and initializes the group's representative $G_{rep} = T_i$. Then, trajectory T_j in $S–G$ that is found to be the nearest to G_{rep} (with respect to a time-tolerant trajectory distance function) is removed from S and inserted into G, and G_{rep} is updated accordingly to the bounding box that encloses T_j and G_{rep} (actually, anonymizing T_j). The algorithm continues until G contains K trajectories. For each anonymized region, the associated time interval is the one that includes the timestamps of the enclosed points. Then, in the next iteration of this step, another group of K trajectories is found, and so on until all trajectories have been assigned to groups of K. In case that (less than K) unmatched points exist, the algorithm suppresses them.

2. *Reconstruction step*: For each group of K trajectories and for each anonymized region, K locations are randomly selected and the corresponding timestamps are also randomly attached from the associated time interval. Then, K trajectories are reconstructed by connecting unique locations from each anonymized region.

The procedure is illustrated in the example of Fig. 8.7. Suppose that the trajectory dataset consists of three trajectories, T_1, T_2, T_3, and $K=3$. So the goal is to reconstruct three trajectories, T_1', T_2', T_3', "similar" to the original ones. In Fig. 8.7a, the algorithm is sketched without the details of the anonymization step: the original dataset (left), which is the input of the anonymization step, is transformed to a sequence of bounding boxes (middle), which is the output of the anonymization step and input of the reconstruction step, out of which the output (fake) trajectories are reconstructed. The details of the anonymization step are unfolded in Fig. 8.7b: first, T_1 is randomly selected and added into empty group G as G_{rep}; then, its nearest trajectory, T_2, is inserted into G and G_{rep} is formed, accordingly, as the sequence of bounding boxes that enclose the respective points of members of G; then, T_3 is added into G, and G_{rep} is reshaped to enclose it.

8.3   Privacy Preserving Mobility Data Querying

In the previous section, the goal was to publish an anonymized version of the original MOD. The main assumption made by those approaches is that, as soon as a trajectory dataset is anonymized, an adversary is no longer able to link the movement of a user with a specific trajectory. However, there exist scenarios where data should stay in-house to the hosting organization and the MOD can only be queried. This is the case when a data holder is not willing or is not allowed, due to regulations, to publish a dataset, even if it is made anonymous. However, assuming that at least part of the data has to become available to possibly untrusted third parties for analysis purposes, a mechanism is needed in order to ensure that no sensitive information will be released during this querying process.

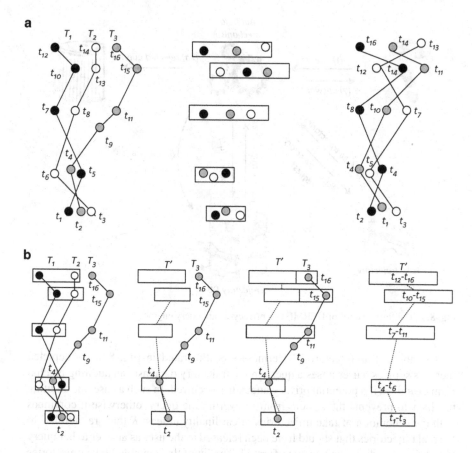

Fig. 8.7 An example of *AWO*: (**a**) the high-level concept, (**b**) the anonymization step in detail

Such a protection mechanism, in the form of a *privacy-aware query engine*, allows subscribed users to gain restricted access to the trajectory dataset in order to perform various analysis tasks while, at the same time, preserving users' privacy from various types of attacks. It should be able to support various types of spatial and spatio-temporal queries. In order to ensure *K*-anonymity, when a user poses a query, the engine retrieves the real trajectories that make the answer set, of cardinality *r*, and, if *r<K*, combines them with *K−r* fake trajectories, by ensuring that the malevolent is not able to distinguish the real from the fake trajectories in the released answer set.

Hermes++ is a privacy-aware query engine that follows, at a large percentage, the above paradigm. It is able to (i) audit queries for trajectory data in order to block potential attacks to users' privacy, (ii) support range, distance, and *k*-nearest neighbor queries, and (iii) preserve user anonymity in answers to queries by (a) returning a set of carefully crafted, realistic fakes trajectories and (b) ensuring that no user-specific sensitive locations are reported as part of the returned trajectories.

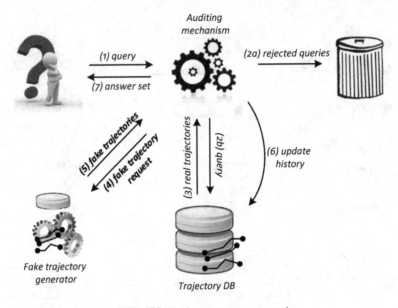

Fig. 8.8 The 'big picture' of HERMES++ privacy-aware query engine

The abstract architecture of Hermes++ is illustrated in Fig. 8.8. As depicted there, as soon as a user poses a query in the trajectory database, an auditing mechanism examines if a potential privacy attack has occurred. In such a case, the auditing mechanism prevents the system from answering the query, otherwise it continues with the generation of fake trajectories of cardinality at least K that are 'similar' to the real trajectories that should have been released to the user as answer to her query if no privacy auditing had been performed. The set of the generated fake trajectories is returned to the user as the answer set and is stored in the MOD for future use (the latter is important for repeatability purposes).

The types of attacks that Hermes++ is able to handle through its auditing mechanism are the following:

- *User identification attack*: the malevolent tries to identify a user by performing ad hoc queries involving overlapping spatiotemporal regions.
- *Sensitive location tracking attack*: the malevolent tries to infer sensitive information by map matching sensitive places of users' trajectory with known locations such as starting and/or ending position.
- *Sequential tracking attack*: the malevolent tries to "follow" a user by performing a set of adjusted queries on nearby regions, in terms of space and time.

In case where the privacy of the users is maintained, the mechanism proceeds to the generation of N realistic fake trajectories, where N is an owner-specified threshold. *Fake Trajectory Generation (FTG)*, the core Hermes++ algorithm, aims at capturing the trend of the result set of real trajectories in order to generate as much realist fake trajectories as possible. FTG extends the *Representative Trajectory Generation (RTG) algorithm* (see Chap. 7) by taking into account the temporal

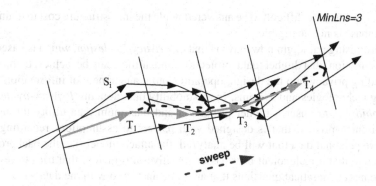

Fig. 8.9 Fake trajectory generation in Hermes++

dimension. Trajectories are first decomposed into segments, which are then filtered in such a way as to keep only those that move towards the same direction. Then, RTG is used so as find their representative trajectory (see the black dotted line in Fig. 8.9), which is swept along its average direction vector so as to modify its two-dimensional segments to realistic two-dimensional and one-dimensional (i.e. along the temporal dimension) segments (see the grey solid line in Fig. 8.9).

8.4 Summary

With the increasing availability of location and mobility data even more privacy threats have been emerged regarding the disclosure of users' sensitive information. In this chapter, we presented several approaches that have been suggested to address the problem of individuals' privacy preservation. In particular, we investigated various privacy-aware techniques and methodologies for location and mobility data whose complex nature imposes special tackling.

Focusing on mobility data, privacy-preserving data sharing approaches adopt the K-anonymity principle in order to sanitize and publish a MOD. They are quite popular techniques since they offer constant data availability in case of analysis purposes. There is no infrastructure cost and are able to model most of releases. However, data need to be anonymized according to pre-specified privacy and utility requirements. Moreover, once an anonymized dataset is released, the data holder has no control over it. On the other hand, privacy-preserving mobility data querying techniques are based on the assumption that data should stay in-house to the hosting organization. Thus, a mechanism should be able to regulate the information that is disclosed to untrusted end users when querying the database. Following this more conservative approach, stronger privacy guarantees are offered and in case of privacy violation, recovery is possible. Moreover, various types of attacks such as user identification, sensitive location and sequential tracking attack can be handled through auditing mechanism. The drawbacks of these approaches correspond to the reduction of data utility since more noise is introduced in the database. More complex queries are difficult to be answered while the infrastructure cost to maintain the database is not negligible.

A promising paradigm nowadays is that of *Privacy-by-design*, which is based on the assumption that higher data protection and utility can be achieved through embedding privacy into the design, operation and management of information processing technologies and systems. In the mobility data domain, *Privacy-by-design for mobility data* ensures higher protection and maximum data utility through a goal-oriented process that is designed with respect to assumptions regarding the sensitive personal data that will be analyzed, the attack model (i.e. the background knowledge that a malevolent user has or sensitive information that tries to reveal) and the target analytical questions that are to be answered with the data.

8.5 Exercises

Ex. 8.1. Collect your own movement data by using your smartphone. The collected GPS log comprises a simple relational database. Annotate each tuple of this database with your activities corresponding to each recorded spatiotemporal point. Sanitize your data by using an off-the-self relational anonymization algorithm. Do you find any risks or potential breaches in your sensitive information? Would you publish this dataset to your favorite social network, say Facebook?

Ex. 8.2. Many trajectory anonymization algorithms base on the results of a clustering algorithm as a preprocessing step. For instance, NWA uses a greedy clustering algorithm. What about the trajectory clustering algorithms discussed in Chap. 7? Could they be used for this purpose? What characteristics are required in order to do so?

Ex. 8.3. Download a dataset from www.chorochronos.org. Apply the NWA and AWO K-anonymization techniques and compare the results. Which technique caused the minimum distortion in the original dataset? How is the distortion affected by the K parameter?

Ex. 8.4. Upon the dataset of the previous exercise, try to exploit on a trajectory sampling technique, such as T-Sampling discussed in Chap. 7, in order to provide anonymization of the dataset. Could sampling be a solution to the trajectory data anonymization problem?

Ex. 8.5. Assume a sanitized (anonymized) version D' of a historical trajectory database D. Your goal is to re-identify trajectories (e.g. based on probabilistic techniques) and approach as much as possible the original D. How would you proceed (as a malevolent user!) to succeed your goal? If it helps, you may assume that you are aware of the anonymization algorithm that produced D' from D.

Ex. 8.6. Hermes++ supports range, distance, and k– nearest neighbor queries on trajectory data. Your goal is to extend an auditing mechanism like that of Hermes++ in order to support path queries (e.g. find objects that pass from a given path on a road network).

8.6   Bibliographical Notes

Snapshot and continuous LBS have been discussed in detail in Gkoulalas-Divanis et al. (2010) and Chow and Mokbel (2011), respectively. The *Center Cloak* and *NN-Cloak* techniques, presented in Sect. 8.1.1, are proposed in Kalnis et al. (2007) while *Casper* technique is proposed in Mokbel et al. (2006). *Interval Cloak* algorithm, proposed by Gruteser and Grunwald (2003), is similar to *Casper* with the difference that is does not deliberate neighboring cells at the same level, but ascents directly to the ancestor level. *Probabilistic Cloaking* is proposed in (Cheng et al. 2006).

   Group-based cloaking for continuous LBS presented in Sect. 8.2.2 is proposed in Chow and Mokbel (2007). *Greedy Cloaking* algorithm (*GCA*), proposed by Pan et al. (2009), is designed to overcome some of the drawbacks of group-based cloaking. Prediction-based cloaking, presented in Sect. 8.1.2, is proposed in Xu and Cai (2008). The *mix-zones* approach has been proposed by Beresford and Stajano (2003), while mix-zones in vehicular networks have been studied by Freudiger et al. (2007) and Palanisamy and Liu (2011). To overcome mix-zones' shortcomings, Zacharouli et al. (2007) proposed an approach that creates dynamically mix-zones on demand.

   Regarding privacy preserving mobility data publishing, the *Never Walk Alone* (*NWA*) technique presented in Sect. 8.2.1 is proposed in Abul et al. (2008). A shortcoming of the *NWA* algorithm is the use of Euclidean distance function in the clustering phase. However, this trajectory distance function is sensitive to outliers and local time swift (see relevant discussion in Chap. 6), while it can only be applied in trajectories synchronized in time. Due to this fact, a preprocessing step aiming to create equivalence classes is required. To overcome this deficiency, an extension called *Wait for Me* (*W4M*) was proposed in Abul et al. (2010), which instead is based on EDR time-tolerant distance function. The preprocessing step is no longer used and during the clustering phase the whole dataset is processed, instead of clustering each equivalence class. The *Always Walk with Others* (*AWO*) technique presented in Sect. 8.2.2 is proposed in Nergiz et al. (2008). Another generalization-based algorithm, called *PPSG*, was proposed by Monreale et al. (2010); the proposed algorithm consists of three phases: (a) generalization, (b) progressive generalization, and (c) trajectory K-anonymization phase. A data perturbation algorithm, known as *Path Confusion*, is proposed in Hoh and Gruteser (2005). When two non-intersecting trajectories are close enough, the algorithm generates a fake crossing in the sanitized dataset so as the adversary can no longer follow the same user with high certainty. Terrovitis and Mamoulis (2008) proposed a suppression-based approach. The dataset is considered as sequences of places visited by users where in each location a transaction has occurred. The basic assumption about adversary's knowledge is that he holds partial information of users' trajectories. The algorithm eradicates the least number of places from a user's trajectory so that the remaining trajectory is *K-ano*nymous with respect to adversary's knowledge.

   The approach of a privacy-aware trajectory query engine was proposed by Gkoulalas-Divanis and Verykios (2008) and was materialized in Hermes++ by

Pelekis et al. (2011, 2012). *Privacy-by-design* is an emerging standard for assuring privacy in the information era (Cavoukian 2012). Its objectives, to guarantee individuals' privacy while gaining a sustainable competitive advantage, are fulfilled through the application of seven foundation principles. *Privacy-by-design for mobility data* was introduced by Monreale et al. (2010, 2011).

References

Abul O, Bonchi F, Nanni M (2008) Never walk alone: uncertainty for anonymity in moving objects databases. In: Proceedings of ICDE

Abul O, Bonchi F, Nanni M (2010) Anonymization of moving objects databases by clustering and perturbation. Inf Syst 35(8):884–910

Beresford AR, Stajano F (2003) Location privacy in pervasive computing. IEEE Pervasive Comput 2(1):46–55

Cavoukian A (2012) Privacy by design [leading edge]. IEEE Technol Soc Mag 31(4):18–19

Cheng R, Zhang Y, Bertino E, Prabhakar S (2006) Preserving user location privacy in mobile data management infrastructures. In: Proceedings of PET

Chow CY, Mokbel M (2007) Enabling private continuous queries for revealed user locations. In: Proceedings of SSTD

Chow CY, Mokbel MF (2011) Trajectory privacy in location-based services and data publication. ACM SIGKDD Explor 13(1):19–29

Freudiger J, Raya M, Félegyházi M, Papadimitratos P (2007) Mix-zones for location privacy in vehicular networks. In: Proceedings of Win-ITS

Gkoulalas-Divanis A, Verykios VS (2008) A privacy–aware trajectory tracking query engine. SIGKDD Explor 10(1):40–49

Gkoulalas-Divanis A, Kalnis P, Verykios VS (2010) Providing k-anonymity in location based services. ACM SIGKDD Explor 12(1):3–10

Gruteser M, Grunwald D (2003) Anonymous usage of location-based services through spatial and temporal cloaking. In: Proceedings of MOBISYS

Hoh B, Gruteser M (2005) Protecting location privacy through path confusion. In: Proceedings of SECURECOMM

Kalnis P, Ghinita G, Mouratidis K, Papadias D (2007) Preventing location-based identity inference in anonymous spatial queries. IEEE Trans Knowl Data Eng 19(12):1719–1733

Mokbel MF, Chow CY, Aref WG (2006) The new Casper: query processing for location services without compromising privacy. In: Proceedings of VLDB

Monreale A, Andrienko G, Andrienko N, Giannotti F, Pedreschi D, Rinzivillo S, Wrobel S (2010) Movement data anonymity through generalization. Trans Data Priv 3(2):91–121

Monreale A, Trasarti R, Pedreschi D, Renso C, Bogorny V (2011) C-safety: a framework for the anonymization of semantic trajectories. Trans Data Priv 4(2):73–101

Nergiz ME, Atzori M, Saygin Y (2008) Towards trajectory anonymization: a generalization-based approach. In: Proceedings of ACM GIS workshop on security and privacy in GIS and LBS

Palanisamy B, Liu L (2011) MobiMix: protecting location privacy with mix-zones over road networks. In: Proceedings of ICDE

Pan X, Meng X, Xu J (2009) Distortion-based anonymity for continuous queries in location-based mobile services. In: Proceedings of GIS

Pelekis N, Gkoulalas-Divanis A, Vodas M, Kopanaki D, Theodoridis Y (2011) Privacy-aware querying over sensitive trajectory data. In: Proceedings of CIKM

Pelekis N, Gkoulalas-Divanis A, Vodas M, Plemenos A, Kopanaki D, Theodoridis Y (2012) Private-HERMES: a benchmark framework for privacy-preserving mobility data querying and mining methods. In: Proceedings of EDBT

Terrovitis M, Mamoulis N (2008) Privacy preservation in the publication of trajectories. In: Proceedings of MDM

Xu T, Cai Y (2008) Exploring historical location data for anonymity preservation in location-based services. In: Proceedings of INFOCOM

Zacharouli P, Gkoulalas-Divanis A, Verykios V (2007) A k-anonymity model for spatiotemporal data. In: Proceedings of STDM

Part IV
Advanced Topics

The map is not the territory.

Alfred Korzybski

Chapter 9
Semantic Aspects on Mobility Data

The bang of mobility data (due to the evolution of positioning devices such as GPS-enabled smartphones and tablets, on-board navigation systems in vehicles, vessels and planes, smart chips embedded in animals, etc.) has an equal share in what is called the BIG DATA era that raises important issues for Moving Object Databases (MOD) and Trajectory Data Warehouses (TDW), which are responsible for the operational and analytical, respectively, processing of moving object trajectories. A reasonable question that arises, is whether we really need all this detailed (i.e. point-by-point) information in order to perform the above processing effectively (i.e. having advanced mobile-aware applications and services in mind). Trying to address this question, during the recent years mobility data are accompanied by *semantic* information (such as diaries filled in manually by citizens for urban transportation research purposes). In a different scenario, semantic information may be inferred by methods taking into account contextual information from the underlying application scenario. Thus, the answer to the above question may be simple: extract and manage (the necessary) semantics from movement and provide services and applications that are built upon them. In this chapter, we first present the background knowledge that allows us to swift the paradigm from raw trajectories to their semantic counterpart, and, subsequently, we study several methods that support a step-by-step methodology towards the reconstruction of the semantically enriched trajectories. The previous reflect the majority of the approaches that have been pursued in the literature, which tackle the raised issues from a conceptual point of view. Then, we go one step further by providing a blueprint of a prototype framework for designing and building real-world semantic-aware MODs and TDWs. Finally, we discuss the semantic aspects of privacy as an orthogonal dimension to the aforementioned techniques.

N. Pelekis and Y. Theodoridis, *Mobility Data Management and Exploration*,
DOI 10.1007/978-1-4939-0392-4_9, © Springer Science+Business Media New York 2014

9.1   From Raw to Semantic Trajectories

Currently, Location-based Services (LBS) or more advanced social networking applications (Facebook Places, Foursquare, Twitter, etc.) make only use of users' current locations, whereas the exploitation of the historic movement (the trajectory) seems to be the next wave to come soon. In order to make use of this historic information to develop novel applications or improve the quality of the existing ones, an approach could be to build upon the results on the mobility data mining domain (i.e. interesting mobility patterns), which operate on raw trajectory data. On the other hand, a parallel line of research that tries to incorporate semantics into the game, comprehending the necessity to take into advantage of such additional information. A semantically-annotated trajectory, in short *semantic trajectory* is an alternative representation of the motion path of a moving object; Fig. 9.1 illustrates the two alternatives: on the one hand (top), a raw trajectory is 'simply' a sequence of time-stamped locations; on the other hand (bottom), a semantic trajectory is a sequence of "interesting information" about transitions from one place to another, information like departure or destination places of interest (POI), roadways, transportation means, activities, etc.

Informally, a semantic trajectory can be defined as a sequence of '*Stop*' and '*Move*' episodes, each with associated meta-data ('*Tags*'), where:

- *Stops* are the parts of the object's trajectory during which the object stays "static" at a place (point or region);
- *Moves* are the parts of the object's trajectory in between two Stops, i.e. where the object is "moving";
- *Tags* are meta-data associated with Stops and Moves.

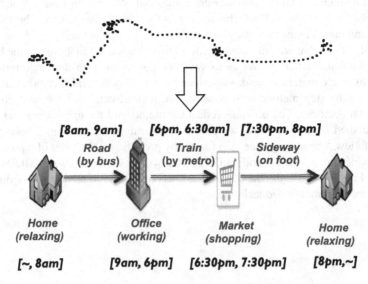

Fig. 9.1 Raw vs. semantic trajectories

It is not only a matter of downscaling the size of a MOD; maintaining semantic trajectories turns to be quite useful in terms of movement analysis (which are expected to be the key issue in next-generation LBS as argued earlier). This is because when movement is perceived as a sequence of (Stop and Move) episodes—e.g. staying home, shopping, moving by bus to workplace, shopping in a shopping mall, etc.—it implies treating mobility data as semantically rich approximations of homogenous fractions of movement, thus allowing better mobility understanding.

The roadmap for semantic trajectory exploitation consists of (a) the transformation of raw trajectories to their semantically enriched counterparts, (b) the efficient data management of resulting semantic trajectories, (c) the emanated analytical and mining opportunities that arise when having this information in our hands, and (d) the unavoidable privacy aspects that have to be addressed. The advances in the above research domains allow us to be optimistic that this paradigm shift may be the new inspiration wave for the mobility research community.

9.2 The Semantic Enrichment Process of Raw Trajectories

Adding application-oriented contextual information to raw trajectories, a process called *semantic trajectory enrichment*, is the key for supporting mobility and behavioral analyses that are of interest to the application at hand. The mobility literature is rich in proposals for the semantic enrichment of raw trajectories, each one built on top of background domain-specific knowledge. A typical trajectory semantic enrichment process receives as input a set of sound raw trajectories (recall the trajectory reconstruction process discussed in Chap. 3), a repository maintaining contextual data and produces as output a set of semantic trajectories. A raw trajectory can be annotated at different levels of detail (i.e. granularities): from the one extreme as a whole (e.g. a trajectory can be classified as "touristic") to the other extreme at point-by-point level (e.g. a $<x, y, t>$ triplet corresponds to a "workplace"). Obviously, annotating the trajectory as a whole may be inefficient in several cases whereas annotating one-by-one the positions of a trajectory generates a large number of repetitive annotations; it sounds more effective to partition the trajectory at meaningful parts (sub-trajectories) that correspond to specific behavior (e.g. "staying at home") or activity (e.g. "leisure"). Therefore, annotation is usually performed at the sub-trajectory (i.e. episode) or, less often, at the trajectory level.

The annotation of episodes requires a pre-processing step: their identification within the trajectory. The process that recognizes the episodes is usually based on a segmentation (i.e. partitioning) method, called *trajectory segmentation*. In the following, we elaborate on trajectory segmentation techniques, focusing on methods that discover Stop episodes (and, as such by deduction, discover Moves), and then, on techniques that annotate the previously discovered episodes.

9.2.1 Trajectory Segmentation and Stop Discovery

Trajectory segmentation is the process of segmenting a trajectory (i.e. a sequence of spatiotemporal positions) to meaningful sub-sequences (i.e. sub-trajectories, hence episodes), according to some application requirements. In detail, each episode is a maximal subsequence such that all of its spatio-temporal positions comply with a given predicate. The predicate usually bears on the spatio-temporal coordinates of the positions. Given this predicate, one can attach additional contextual information (e.g. POIs, activities etc.) to each episode and thus enrich the trajectory with semantic information. As expected from the above discussion, the most popular segmentation criterion is whether a moving object is stationary or not. Based on this, different algorithms have been proposed to discover Stops within a raw trajectory.

A first approach simply associates Stops with the parts of a raw trajectory where there is either absence of signal of the positioning device (e.g. the GPS is off), or the velocity of the moving object is zero for a given temporal interval (e.g. the car is parked). The weakness of this approach lies on the possibility that an actual Stop may not be identified, as due to signal errors, the measured speed is slightly above zero. A straightforward extension to resolve this issue is to use speed (e.g. less than 1 km/h), area coverage (e.g. less than 50 m^2), and temporal duration thresholds (e.g. more than 5 min) to imply stillness. Of course, all these thresholds are application-oriented, being also sensitive to the 'interestingness' of a Stop; for instance, it is natural that the more time is spent in a place, the more important is the place for the moving person.

Stop discovery may gain from combining contextual geographic information. For instance, a Stop is discovered when the trajectory intersects the geometry of an 'interesting' place (from a set of predefined POI's) and the duration of intersection is above a given temporal duration threshold. Alternatively, by detecting dense areas of the trajectory points, using e.g. a density-based point clustering method (like DBSCAN), one may identify clusters, which correspond to the 'slower' parts of the trajectory, the potential Stops. Then, these candidates are mapped to POI's, considering the geography of the region, resulting in the actual Stops. The two approaches sketched above are followed by SMoT and its extension, CB-SMoT technique, respectively. They can be classified as density-based Stop discovery techniques with the main idea illustrated in Fig. 9.2a.

Among movement features (velocity, acceleration, heading, etc.) velocity seems to be the most related to the Stop discovery problem. When segmenting a trajectory with respect to velocity, a speed threshold can be set, so as, when the instant speed of a point is below the threshold, then it is assumed to be part of a Stop, otherwise it is assumed to be part of a Move episode. Also, a minimum stop time parameter can be set in order to avoid short-term congestion (e.g. in traffic lights) to be considered as Stop; the idea is illustrated in Fig. 9.2b.

The previously discussed trajectory segmentation (via Stop discovery) approaches are mainly offline, which is not enough for dynamic, real world applications that require service in an online and continuous fashion, since mobility data usually

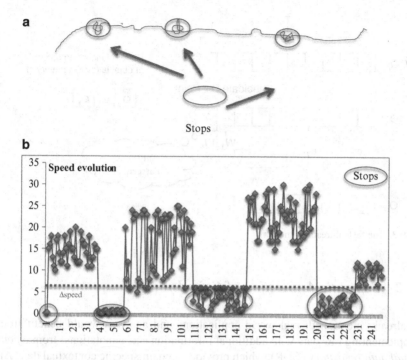

Fig. 9.2 The Stop discovery problem: (**a**) point density-based versus (**b**) velocity-based stop identification.

arrive as streams. Moreover, online semantic trajectory construction can be useful in many traffic monitoring scenarios where authorities make use of queries of the form: *"Report every τ secs the movement and driving behavior of the objects within area A during the last T minutes"*. SeTraStream is a real-time platform that can progressively process raw mobility data arriving within a restricted time window and compute semantic-aware trajectories online. The workflow of SeTraStream is illustrated in Fig. 9.3.

More specifically, upon the receipt of a batch containing the status updates including the movement attributes of object O_i at timestamp t, which are described by a d-dimensional vector, called *Movement Feature Vector* (MFV), at different timestamps in $τ$, a cleaning and smoothing technique is applied on it (step 1). Consequently, a compression method (step 2) is applied on the batch considering the MFV characteristics. Finally (step 3), a matrix from the MFV is formed and the batch is buffered until it is processed at the SeTraStream's segmentation stage. During the segmentation stage (left part of Fig. 9.3), a previously buffered batch is dequeued and compared with other batches' feature matrices in O_i's window. SeTraStream seeks both for short and long term changes in O_i's movement pattern, and identifies an episode whenever feature matrices are found to be dissimilar based on correlation function and a specified division threshold $σ$.

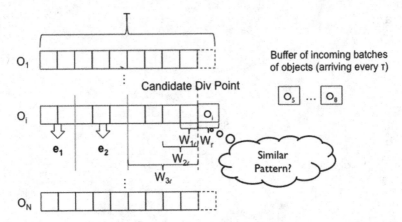

Fig. 9.3 The SeTraStream framework

9.2.2 Semantic Annotation of Episodes

As already mentioned, real world applications and analyses would benefit from complementing raw data with additional information (i.e. annotations) from a *contextual data repository* (CDR), which provides domain-specific contextual data. An annotation value is either a literal (e.g. the term "shopping" is a potential value for a *ActivityMeans* annotation), or a link to an object in the CDR (e.g. the key "shopping_mall_X" in a Shopping_Malls table), or even a combination of the two. Annotation values may be recorded by witnesses (e.g. drivers stuck in a traffic congestion) or by sensors (e.g. heart rate measured by a smartphone). They may be computed from raw data (e.g. distance from a POI), extracted from contextual data (e.g. registered events at a POI), and inferred by reasoning (e.g. an activity may be inferred from the category of a POI and the past habits of a user).

The implicit annotation of an episode as Stop or Move via a segmentation method is called *defining annotation*. The annotation of the activity of a user in a Stop or Move episode, or the transportation mode during a Move is called *episode annotation*. The annotation at the trajectory level is called *trajectory annotation*. In the following paragraphs we will survey annotation methods that take place at the episode and trajectory level, assuming that a segmentation method has already been applied in order to discover Stop/Move episodes.

In order to capture *episode annotation*, a straightforward solution is to base on the intrinsic idea of the utilized trajectory segmentation method. For instance, a Stop can be annotated with a POI that is "the closest to" the Stop (i.e. with respect to a proximity criterion) or "the most likely" to the Stop (i.e. with respect to both a proximity and a compliance criterion; an 8-h Stop better fits in a "workplace" than a "dining in a restaurant" activity).

On the other hand, a popular criterion for annotating a Move episode is the means of transportation used by the moving object. The typical approach is the characterization of each transportation mode, by considering parameters, such as speed

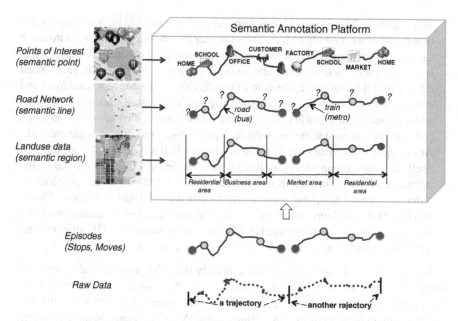

Fig. 9.4 Trajectory annotation according to SeMiTri framework

(e.g. walking speed is less than 5 km/h), motion continuity (e.g. buses make frequent stops while taxis do not), route constraints (e.g. buses and trams use move on pre-defined routes), etc. Following this line, in a first step we could detect the positions where movement switches between walking and non-walking segments while in a second step, we could refine the non-walking segments into segments characterized by other transportation modes (bicycle, bus, car, etc.).

SeMiTri is a complete three-layers framework aiming to support the semantic annotation of raw trajectories with contextual data coming from application and geographical sources. More specifically, after the Stop detection step, the approach annotates Stop and Move episodes with geographical objects (points, lines, regions) using spatial joins with respect to directional, proximity, and topological spatial relationships.

The overall SeMiTri framework is illustrated in Fig. 9.4. The three levels of the architecture operate as follows:

– Annotation with geographical regions relies on the identification of Regions of Interest (ROIs). There ROIs may either be regions of any shape, or cells of a grid tessellation. For the first, OpenStreetMaps regions are used, which are then joined spatially with the trajectories to identify topological correlations. After intersected regions have been found, continuous GPS points are grouped, the entering and leaving timepoints is calculated and a tuple $<region_i, t_{in}, t_{out}, region_{type}>$ is formed. Consecutive tuples with the same region type are merged and, as such, a trajectory is transformed to a sequence of regions.
– The second layer annotates trajectories with semantic lines. The process takes place in two steps: First a global map-matching technique is used to link road

segments with Move episodes and then a classifier is used to infer the transportation mode of the moving object. The classifier uses features separated into three categories: movement's features (i.e. episodes' travelled distance, duration, average speed etc.), road network features (i.e. road type, road length, road speed etc.), and POI features (i.e. the type of the POI where the Move started from and ended to). For each map-matched road segment the feature vector is calculated and is passed to the classifier model. The classifier output the transportation mode and consecutive road segments with the same transportation modes are merged.

- At the third layer, Stops are annotated with the POI's type rather than the POI itself, as dense urban areas can have several different POIs, which makes the inference of the exact POI from imprecise location records a hard problem. In order to succeed this, the approach is modeled as a Hidden Markov Model (HMM) problem. More specifically, after the raw trajectory is transformed to a sequence of stops, this forms the HMM observation, while the POI categories are the hidden states that we are interested in, which come from the exact POI.

Instead of identifying the annotations directly from the geographical objects, an alternative approach is to annotate episodes with the activities taking place (e.g. "shopping" instead of "shopping_mall_X"). This inference of activities utilizes their particular characteristics. For instance, most POI's are related to more than one activity and this should be taken into consideration (e.g. "working" could be also used for a stop at "shopping_mall_X"); Moreover, certain activities have specific periodicity and/or temporal variability (e.g. "sleeping" is usually performed during and every night). Intuitively, the inference would be more effective if activity workflows are taken into account as constraints, (e.g. a "bring children to school" activity may be followed by a reverse "get children from school" activity). Finally, additional constraints such as the number of "work" activities per day should be part of the inference mechanism.

A rather more difficult task is to annotate the whole trajectory of a moving object. To do so, an approach could be to aggregate the various annotations into a single annotation that best describes the overall trajectory. For instance, the trajectories of a user going every day to a sporting center for 3 h implies annotating them as "athletics". Another approach to succeed such holistic annotation is to identify 'trips' (i.e. sequences of episodes where intermediate episodes actually pertain to a final goal) and use only the activity of the last episode of the trip, or to use common sense if-then rules to annotate the trajectory with the most probable global activity, which is derived by the frequency of the activities at Stop episodes.

9.3   Semantic Trajectory Data Management

It is clear that raw and semantic trajectory databases (TD and STD, respectively) do not have much in common in their representation. As a result, an existing MOD engine, such as the state-of-the-art Hermes and Secondo, cannot be used as-is in order to handle semantic trajectory databases. As already discussed in Chaps. 4 and 5, efficient TD support is already out there. On the other hand, efficient STD support is missing, which raises very challenging tasks for research (how to store, index, and query STDs). Consider, for instance, the following queries:

- *"Search for people who follow the home—office—home pattern every weekday"*;
- *"Search for people who cross the city center on their way from office back to home"*;
- *"Search for people who make long trips (e.g. more than 20 km) on their way from home to office without include intermediate stops"*.

Such queries are innovative and cannot be handled effectively and efficiently by existing approaches and corresponding MOD engines.

In the same line, designing and efficiently feeding a corresponding semantic trajectory data cube for multi-dimensional analysis is a natural direction to follow. Thus it would be able to support analysis of type *"when, where and why moving objects of a specific profile stop?"* and *"when, how and where from/to moving objects of a specific profile move?"*, focusing on Stop and Move episodes, respectively.

Moreover, in real applications two distinct repositories are foreseen: a *trajectory database* (TD) consisting of raw trajectories and a *semantic trajectory database* (STD) consisting of semantic trajectories. These two repositories should (a) be efficiently managed and (b) collaborate each other in order for the desired functionality to be offered, as illustrated in Fig. 9.5.

In the following sections, we provide a blueprint for designing and developing such a framework. More specifically, we present a datatype system for the representation of semantic trajectories and provide details for the development of such a STD into an extensible DBMS architecture, following the MOD engine paradigm. Then, we discuss baseline access methods for the hybrid indexing of both the spatio-temporal as well as the semantic component of such data, so as to support the previously mentioned query examples.

9.3.1 A Datatype System for Semantic Trajectories

In order to store semantic trajectories, defined according to the assumptions of Sect. 9.1, into an extensible DBMS, we need to define appropriate datatypes that, on the one hand, support the adopted conceptual model, and, on the other hand, allow for efficient implementation. Adopting the usual definition of a (raw) trajectory datatype τ of a moving object as a triple ($o\text{-}id$, $traj\text{-}id$, T), where $o\text{-}id$ ($traj\text{-}id$) is the identifier of a moving object (the specific trajectory of a moving object, respectively) and T is a 3-dimensional polyline representing its movement. In turn, the concept of a (raw) sub-trajectory τ' of a (raw) trajectory τ may be represented by the same datatype. Thus, a sub-trajectory is similarly defined as a triple ($o\text{-}id$, $traj\text{-}id$, $subtraj\text{-}id$, T'), where $subtraj\text{-}id$ is the identifier of the specific sub-trajectory of the moving object, and T' is the portion of T between two timestamps, t_i and t_j, $t_i \le t_j$. More specifically:

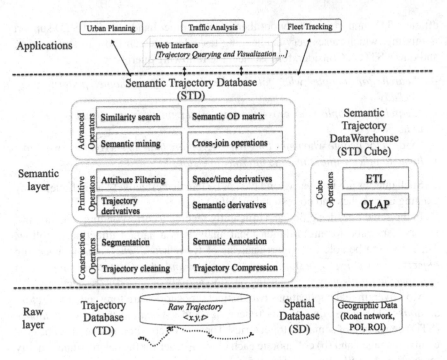

Fig. 9.5 Big view of a semantic trajectory DB/DW infrastructure

- An **episode** e of a semantic trajectory corresponds to a (raw) sub-trajectory τ' and is defined as a tuple (*defineTag, MBB, episodeTag, activityTag, T-link*), where *defineTag* is a flag that defines the episode as Stop or Move *MBB* is a tuple (*MBR, t_{start}, t_{end}*) corresponding to the 3-dimensional approximation of τ', where *MBR* is the enclosing rectangle of the spatial projection of τ' in 2-dimensional plane and [t_{start}, t_{end}] is the temporal projection of τ' in 1-dimensional timeline, *episodeTag* and *activityTag* hold semantic (i.e. textual) information about Stops or Moves, and *T-link* is a link to the spatio-temporal representation of τ'.
- A **semantic trajectory** τ_{sem} of a moving object corresponds to a (raw) trajectory τ and is defined as a triple (*o-id, semtraj-id, T_{sem}*), where *o-id* (*semtraj-id*) is the identifier of a moving object (the semantic trajectory of a moving object, respectively) and T_{sem} is a sequence of episodes, {e_1, ..., e_n} belonging to τ and being ordered in time, i.e. $e_i[t_{end}] \le e_{i+1}[t_{start}]$, $1 \le i < n$.

In order to support such query functionality via a query language, similarly to the approach followed in raw TDs, one should define a set of object methods for the datatypes previously defined, as well as a set of operators to be applied in relations having columns of such datatypes. Below, we list a number of primitive methods/operators on STDs, which by no means are exhaustive (in contrast, our purpose is for them to be used as a food for thought).

Primitive methods for the *episode* (and, accordingly, the *semantic trajectory*) datatype may include:

- number *duration* () : returns the duration (in sec) of an episode
- number *length* () : returns the length (in m) of an episode
- geometry *PAA* () : returns the Potential Area of Activity (PAA) of an episode
- number *avg-speed* () : returns the average speed (in m/sec) of an episode
- boolean *intersects* (MBB *b*): returns true or false whether the MBB of an episode intersects a spatio-temporal box *b* or not.

Some additional object methods for the *semantic trajectory* datatype, which facilitate filtering w.r.t. any conjunction of spatial/temporal/semantic conditions:

- number *num_of_episodes* (string *tag,* string *distinct*) : This function takes as input a *tag* string of the form *"tag1+tag2+....+tagn"* (i.e. implying a concatenated set of (sub-)strings), and a *distinct* string that can be either "yes" or "no" (i.e. implying a boolean flag). It returns the number of episodes (distinct or not, depending on the use of the flag) of the semantic trajectory that includes tags LIKE the given ones. In this case, "LIKE" implies pattern-matching per input *tag,* conjucted by an OR logical operation.
- set[episode] *episodes_with* (string *tag*): This function takes as input a *tag* as previously and instead of counting the matched episodes, it returns a collection of them. In case where an episode matches multiple times with some input tags, this is returned only once.
- sem_trajectory *confined_in* (geometry *g,* timeperiod *p,* string *tag*): returns the part of a semantic trajectory (a sem_trajectory object) that includes the episodes of the semantic trajectory that (a) spatially overlap with *g,* (b) temporally intersect with *p,* and (c) matches textually with *tag.*

Using the above operators, several (range, nearest neighbor, similarity, etc.) queries can be designed over a semantic trajectory database. Recalling the architecture illustrated in Fig. 9.5, interesting types of queries are classified as follows:

- Q1 type: *raw trajectory queries,* which are those involving TD;
- Q2 type: *semantic trajectory queries,* which are those involving STD;
- Q3 type: *cross-over semantic trajectory queries,* which are those involving both TD and STD.

For instance, *"Search for people who crossed a region (e.g. park X) at night"* is a typical Q1 type query. As already discussed (see Chap. 4), such coordinate- (range, nearest-neighbor, etc.) or trajectory-based queries (enter, cross, etc.) have been extensively studied in the MOD literature. On the other hand, *"Search for people who follow the pattern home—office—home every weekday"* or *"Search for people who cross the city center on their way from office back to home",* Q2 and Q3 type query, respectively, are innovative and they cannot be considered as straightforward variations of Q1 type queries. Especially for Q3 type (cross-over) queries, they can be technically supported assuming *T-link* in the definition of episode datatype (see above), but it is clear that interesting query optimization problems arise.

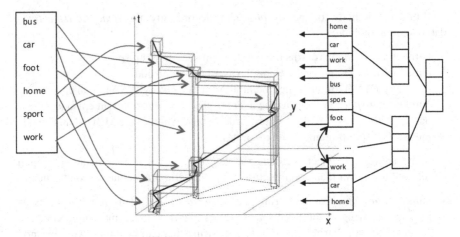

Fig. 9.6 The main idea of STB-tree: a TB-tree enhanced with textual information (*right*) along with an inverted file on tags (*left*)

Further examples of the above foreseen functionality are demonstrated in Chap. 12, where a hands-on analysis based on such a STD framework is presented.

9.3.2 Indexing Semantic-Aware Trajectory Databases

The issue of indexing an STD raises interesting challenges, which have not been studied so far in the literature. The key observation that reveals the inherent difficulty of this task is that one should be able to manage both the spatio-temporal as well as the semantic (i.e. textual) component of semantic trajectories. As a food for thought, we present a baseline access method for the organization of semantic trajectories in extensible ORDBMS.

In particular, in order to index such complex objects one baseline solution could be the use of spatio-textual indexing structures used in geographic information retrieval. However, structures used in this field (i.e. R-trees enhanced with textual search capabilities) ignore the time dimension, as such suffering from the same shortcomings that R-trees have when indexing mobility data. A different alternative would be to use a data structure tailored for mobility data, such as the well-known TB-tree, accompanied with a text index, such as an inverted file. The idea of such a hybrid access method for semantic trajectories, called *Semantic Trajectory Bundle Tree* (STB-tree) is illustrated in Fig. 9.6.

The intuition for choosing TB-tree as the base for STB-tree is its trajectory preservation property, so as each leaf node keeps only those episodes that belong to the same semantic trajectory and the index is organized according to episode MBBs. The basic difference from the original TB-tree is that a (semantic) trajectory is represented as a sequence of episode MBBs instead of 3-dimensional segment MBBs. As such, the procedure for insertion is similar to TB-tree, however it is obvious that the resulted tree is smaller in size. Although the STB-tree is still built upon spatio-temporal information (episode MBBs), it also maintains the textual information for each episode on the corresponding entry for that episode on the leaf node. This is quite useful as it allows combined queries with spatio-temporal and textual constraints to be resolved.

The above discussion prescribes that queries' resolution can take advantage of a filtering step on the spatio-temporal constraints of the query. Then a refinement can take place by using the textual constraints. On the same line, the STB-tree utilizes an inverted file built on the textual terms of the tags of the episodes, so as to enable the reverse filter-refinement process (i.e. first filter by textual terms by using the inverted file and then refine by the spatio-temporal constraints). This is facilitated by allowing each textual term of the inverted file to point to specific entries in the leaf nodes of the tree structure. Obviously, the question of which of the two alternative filter-refinement directions one should follow, is an interesting optimization problem.

9.4 Semantic Trajectory Data Exploration

The data cube paradigm has been utilized to support spatial and (raw) trajectory DWs, involving spatial/temporal versus thematic (alphanumerical) dimensions and spatial/spatio-temporal versus numerical measures; obviously, an effective data warehouse for semantic trajectories inherits from the above. Moreover, mining semantic trajectories is challenging; we discuss some primitive ideas for performing clustering over such complex data objects.

9.4.1 Semantic-Aware Trajectory Data Warehouses

Recalling the scheme of the (raw) trajectory data warehouse (TDW) in Chap. 6, one could extend it and design a *semantic trajectory cube* (STC) as a constellation scheme consisting of at least five dimensions (corresponding to space, time, user profile, STOP-type activity, and MOVE-type activity, respectively) and two fact tables (i.e. a STOPS-fact and a MOVES-fact table). Intuitively, this approach allows the support of the following kinds of analysis:

- *STOPS-Fact* (similar to spatial data cube): who made a stop? when and where? what did she do during her stop?
- *MOVES-Fact* (similar to raw trajectory data cube): who made a movement? when and where from/to? How did she move and what did she do during her motion?

This example STC is illustrated (in relational format) in Fig. 9.7. Let us explore in more detail the above scheme focusing on the measures of the two fact tables, which are the ones that reveal the analytical power of a data cube. To start with, we

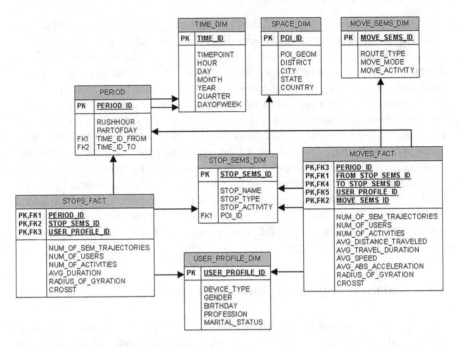

Fig. 9.7 A semantic trajectory data cube

define the cells of the two base cuboids, bc_{stop} and bc_{move}. On the one hand, bc_{stop} is composed of three dimensions over Stop episodes: (i) temporal dimension, (ii) spatially-related semantics, and (iii) user profile; this is illustrated in Fig. 9.7 through the triple <period_id, stop_sems_id, user_profile_id> of foreign keys to the respective dimension tables that stands as primary key of Stops_Fact table. On the other hand, bc_{move} is composed of five dimensions over Move episodes: (i) temporal dimension, (ii) spatially-related semantics focusing on the departure Stop, (iii) spatially-related semantics focusing on the destination Stop, (iv) user profile, and (v) movement-related semantics; in turn, this is illustrated in Fig. 9.7 through the quintuple <period_id, from_stop_sems_id, to_stop_sems_id, user_profile_id, move_sems_id> of foreign keys to the respective dimension tables that stands as primary key of Moves_Fact table.

Note that in both fact tables, the spatial dimension is at the level of POI, while it is modeled via the associated semantics of the POI, as a snowflake scheme. Looking the spatial dimension of the STC at the POI level instead of at a higher region level (as in the case of raw TDW, where the base cuboid is defined by a spatial tessellation) raises interesting computation challenges for the ETL process.

Moreover, Fig. 9.7 presents a representative set of measures of the STC that capture interesting properties of the source STD in an aggregate way. Like in raw TDWs, *num-of-sem-trajectories* is the most essential measure, upon which other measures can be defined. Note that this measure is also subject to the *distinct count problem* discussed for raw trajectories in Chap. 6 and could possibly find similar solutions.

Given the above measure one may define several others. In order to correlate the discussion of the required ETL process, which is in charge of calculating measures in base cells of a data cube, with the query functionality in STD, let us discuss two indicative measures, one for each fact table in our constellation scheme. For instance, *avg-duration* on Stops_Fact table is defined as the average time spent by semantic trajectories inside a bc_{stop}. This is calculated to be the sum of lifespans of the sub-trajectories that correspond to Stop episodes, which are restricted inside bc_{stop} over the number of involved semantic trajectories. This is illustrated formally in the following equation:

$$avg\text{-}duration\left(bc_{stop}\right) = \frac{\sum_{ep_i \in bc_{stop}} duration\left(ep_i\right)}{num\text{-}of\text{-}sem\text{-}trajectories\left(bc_{stop}\right)}$$

where one can note that the corresponding query that supports the corresponding ETL task is a Q3 query type. This is because in some cases we should look back to the raw MOD (via the *T-link* pointer), in order to appropriately restrict the sub-trajectory of the episode.

Similarly, *avg-distance-traveled* measure on Moves_Fact table, defined as:

$$avg\text{-}distance\text{-}traveled\left(bc_{move}\right) = \frac{\sum_{ep_i \in bc_{move}} length\left(ep_i\right)}{num\text{-}of\text{-}sem\text{-}trajectories\left(bc_{move}\right)}$$

corresponds to a Q3 type query, where *length(ep_i)* is the length of the Move episodes confined in bc_{move}.

9.4.2 Mining Semantic Trajectory Databases

In Chap. 7, several methods for mining raw trajectory databases were discussed. Clearly, semantic trajectory mining poses even more interesting challenges, as it is (actually) closer to what an analyst would envision when working with mobility data. For instance, focusing on clustering techniques, it is far more useful to discover a pattern like *"this group of students (coming from different origins) stopped at piazza P participating in a protest for an hour; then walked to café C where they stayed for half an hour; and finally they went by tram T to bar B where they partied until midnight"*, than identifying a pure spatio-temporal pattern (e.g. T-pattern) that lacks semantic information. At the moment, tackling such challenges is done by first applying mobility mining techniques to raw trajectories and then following a semantic enrichment process on the discovered patterns. However, it is interesting to discuss what one can do in order to natively mine semantic trajectory objects.

Focusing on clustering, it is natural to look for a semantic trajectory similarity function as the main building block of the clustering process. To do so, we could follow one of at least two directions. The first bases on the utilization of a feature extraction process that maps semantic trajectories into vectors in an appropriate multi-dimensional feature space. Consider, for example, the following derived properties for each Move episode of a semantic trajectory: (i) the distance (or the area) covered by the moving object during the episode, (ii) the duration of the episode, (iii) the average speed recorded during the episode, (iv) the type of transportation means, (v) the type of departure POI, (vi) the type of destination POI, etc. The (dis-)similarity between two semantic trajectories is then defined as their distance in the multi-dimensional feature space; the closer their feature vectors are located, the more similar the semantic trajectories are.

In a second direction, the (dis-)similarity between two semantic trajectories can be defined by an aggregative function that unifies two similarity functions; one applicable to the spatio-temporal component of semantic trajectories (recall, for instance, the plenty of raw trajectory similarity functions discussed in Chap. 6), and the second that quantifies the semantic (i.e. textual) similarity. Such complex similarity functions can be inspired by spatio-textual approaches employed in geographic information retrieval. One may think of two different similarity functions, one applicable to episodes and the other applicable to whole semantic trajectories. Having these functions in hand, we could employ any of the approaches already successfully applied to raw trajectories (e.g. compute the distance matrix and forward it to a hierarchical clustering algorithm). Using the first function we could discover groups of persons stopping at a region for the same purpose, while with the second we would be able to identify patterns like the group of students discussed earlier.

9.5 Semantic Aspects of Privacy

The need of capturing richer representations of individual's daily movement has led researchers to add conceptual knowledge into application analysis processes. This transition from raw trajectory data to semantically-enriched trajectories emerges even greater privacy violation threats. Consider, for example, an individual visiting a place such as a mental health clinic every Friday evening. It is obvious that such POI contains private information that needs to be protected when publishing mobility data. This type of location is considered as sensitive and the corresponding Stop episode in a semantic trajectory as a *sensitive stop*. Below, we discuss various approaches that tackle such semantic aspects of privacy, in the context of LBS and semantic trajectory databases.

9.5.1 LBS for Sensitive Semantic Locations

There are cases where an adversary is aware of the semantic whereabouts of a territory. In such cases, the additional semantic knowledge might lead to the extraction of private personal information. Thus, the main goal is to protect an LBS user's location when it is a sensitive stop. Similar to privacy in context of snapshot and continuous queries for LBSs (recall the respective discussion in Chap. 8), a solution could be a generalized cloaked region containing at least K users for each sensitive location. But what if all K users lie within the same (sensitive) location? A solution (location l-diversity) suggests that a cloaked region should contain not only at least K users, but also at least l different POIs. The degree of privacy is proportional to l. However, based on the assumption that a malevolent user has background knowledge regarding the probability distributions of the POIs, *semantic location identification attack* might arise. For example, consider that the generalized cloaked region encloses two POIs, a mental health clinic and a school; if the LBS request was sent late in the evening, then it is natural that the malevolent will bet for the clinic.

To prevent the aforementioned type of attack, a semantic location cloaking approach, based on the assumption that different users consider different locations, works as follows: the entire area is divided into equally-sized cells; each cell contains sensitive and non-sensitive places with their corresponding position probabilities; each user is able to determine her privacy profile regarding which POIs are sensitive with the corresponding privacy degree. Then, semantic location cloaking is performed. The set of *cloaked regions* (CR) muffles the sensitive locations and is formulated irrespectively of the users' actual locations based on users' privacy profiles with respect to adversary's background knowledge. A CR should contain not only sensitive but also locations of different types. When a user requests for a LBS, the user's position is examined and if it corresponds to a CR then the CR is returned, otherwise the exact user's position is returned. A so-called *strongly CR* has the following properties: (a) satisfies users' privacy profile specifications, (b) fulfills minimal disclosure requirement (i.e. the probability of linking a user with a sensitive place does not exceed a threshold based on her privacy profile), and (c) contains at least one non-sensitive place.

9.5.2 Privacy in Semantic Trajectory Databases

By applying K-anonymity in semantic trajectories, it is not always possible to control the protection of individuals' privacy. Consider, for example, the case where K trajectories in the STD stop at a clinic, which is considered as a sensitive stop. In this case, if an attacker combines background knowledge, e.g. she is aware of the presence of a specific user in the group of K trajectories, then she can easily infer that the user for sure stops at this sensitive place. An idea, followed by the C-safe Anonymization of Semantic Trajectories (*CAST*) algorithm, proposes to construct a "sanitized" (*c-safe* in CAST terminology) version of a semantic trajectory dataset. In particular, a generalization process is applied on the semantic (rather than the spatio-temporal) level, where sensitive stops are replaced with more abstracted concepts. The approach takes into account that a specific place can be sensitive for a user, but this in not necessarily true for all users. For example, a clinic may be sensitive place for a patient while this is not true for an employee working there. Therefore, quasi-identifiers are regarded as sequences of places (e.g. restaurant, museum, shopping mall, cinema, etc.), where in case of linking with external information they might lead to unique user identification. Hence, place taxonomy is used (by a domain expert), in order to represent the sensitive and non-sensitive places in a hierarchy.

Moreover, the approach assumes that an adversary has gained access to the database and is aware of (i) the utilized anonymization process, (ii) the place taxonomy, (iii) the presence of a user in the dataset, and (iv) the quasi-identifier sequence of visited places. The final goal of this methodology is to publish a *c-safe* version of the original data. The attacker creates a set of candidate semantic trajectories containing the sequence of quasi-identifying places and tries to deduce any other sensitive places related to the specific user, based on the aforementioned knowledge. C denotes the maximum probability that an attacker, given the set of quasi-identifier places, is not be able to infer any other sensitive place of the user. If a quasi-identifier sequence contains sensitive places and the disclosure probability of the given dataset c is less than a given threshold, the sensitive places are generalized. The utility of the data is also maintained in terms of information loss by controlling the cost of generalization process.

9.6   Summary

Modeling, management and knowledge discovery aspects on (raw) trajectory data have been exhaustively studied in the past two decades, including plenty of algorithms and systems, spread from data management to data mining. On the other hand, semantic aspects of mobility data are a relatively new entry in the research agenda. For the moment, research has focused mainly on conceptual modeling approaches of semantic trajectories and on semantic enrichment techniques of raw data, so as to instantiate the proposed models. However, semantic mobility database management and exploration is still in its infancy. This chapter has presented the most representative methodologies proposed so far, while it has discussed baseline methods w.r.t. management and mining such data, also foreseeing some research directions that will very soon be considered as real requirements.

9.7   Exercises

Ex. 9.1. Instead of using the CB-SMoT Stop detection approach, how would you adapt a hierarchical clustering algorithm to identify Stops, which will also be represented at different scales (i.e. hierarchies)?

Ex. 9.2. Consider a region where a moving object Stops (i.e. the MBR of the raw sub-trajectory of the discovered episode). Assuming that there are many POIs inside this region, how would you quantify the corresponding probabilities that the user stopped at these POIs? If you had knowledge of the profile or the habits of the user, how would you adapt them?

Ex. 9.3. Given the methods and operators for SMDs presented in Sect. 9.3, write SQL-like commands to resolve the following queries:

(a) Search for persons that start from home and go to work by bicycle.
(b) For those people that stop for more than 1 h at a super market on their way to home, provide the histogram of the origins of their semantic trajectories.
(c) How much time do mothers spend for bring-get activities of their children at school, considering only home—school—home trips?

Ex. 9.4. Given the SMC of Fig. 9.7 and a partitioning of the geographical space (e.g. by a regular grid), compute the origin-destination (OD) matrix by using the already aggregated data instead of querying the original SMD.

Ex. 9.5. Consider the paradigm of open, linked data (e.g. http://linkedgeodata.org/). How would you model semantic trajectories as triples, so as to store them as an RDF graph? Think of a semantic enrichment process that is driven by (geo-) SPARQL queries to such open data.

9.8 Bibliographical Notes

Many researchers have been challenged by the enrichment of mobility data with semantic information. This line of research has started by several proposals of conceptual models for semantic trajectories (Spaccapietra et al. 2008; Bogorny et al. 2009; Spinsanti et al. 2010; Spaccapietra and Parent 2011; Bogorny et al. 2013). Such conceptual models have recently been accompanied by several methods for reconstructing semantic trajectories from raw data (Yan et al. 2010; Zheng et al. 2010; Yan et al. 2011a; Yan et al. 2011b). In Parent et al. (2013), the interested reader may find an extensive survey of relevant models and techniques.

In Ashbrook and Starner (2003); Krumm and Horvitz (2006); Andrienko et al. (2007); Rocha et al. (2010), the authors have proposed several threshold-based techniques to segment trajectories into parts, aiming at identifying Stops. The clustering-based Stop detection SMoT method was proposed by Alvares et al. (2007), while its extension CB-SMoT was proposed by Palma et al. (2008). A shortcoming of CB-SMoT is on the discovery of Stops after motions with different speeds, e.g. different transportation modes. This is due to the reliance on a global parameter that defines the size of the space to compute the speed in a certain area. To resolve this, the so-called TrajDBSCAN algorithm, proposed by Yan (2011), uses another extension of the DBSCAN clustering algorithm by introducing additional time duration constrain to the discovered clusters. A similar approach has been also followed to extend the OPTICS clustering algorithm by redefining the core and reachability distances so time is taken into account. The advantage of using the so-called T-OPTICS algorithm, proposed by Zimmermann et al. (2009), is that the algorithm behaves better with trajectories with diverse sampling rates (and as such diverse spatial density) and that this approach is interactive due to the utilization of the reachability plot. Liao et al. (2005) used a Gaussian mixture model based on speed, in order to create a new episode whenever a switch in the speed range (i.e. walking, high speed and low speed) is detected. A similar approach is developed in Zheng et al. (2010) where decision tree inference is used.

The SeMiTri framework for semantic trajectory reconstruction presented in Sect. 8.2 was proposed by Yan et al. (2011a, 2012). In certain application domains (e.g. ecology) the annotation with activities is performed manually by the scientists in the field, selecting from a predefined set (Cagnacci et al. 2010). Guc et al. (2008), Spinsanti et al. (2010), and Renso et al. (2012) have proposed methods to annotate the whole trajectory of a moving object instead of annotating specific episodes.

For geographic information retrieval techniques that manage spatio-textual data, the reader is referred to Wu et al. (2012), where several interested related works can inspire research towards the management and analyses of semantic trajectories, in the context that semantics are represented by textual data.

Semantic location cloaking presented in Sect. 9.5 was proposed by Damiani et al. (2011), while it was adapted in the context of a road network by Yigitoglu et al. (2012). The CAST semantic-aware anonymization algorithm was proposed by Monreale et al. (2011).

References

Alvares LO, Bogorny V, Kuijpers B, de Macedo JAF, Moelans B, Vaisman A (2007) A model for enriching trajectories with semantic geographical information. In: Proceedings of GIS

Andrienko G, Andrienko N, Wrobel S (2007) Visual analytics tools for analysis of movement data. ACM SIGKDD Explorations 9(2):38–46

Ashbrook D, Starner T (2003) Using GPS to learn significant locations and predict movement across multiple users. Personal Ubiquitous Computing 7:275–286

Bogorny V, Kuijpers B, Alvares LO (2009) ST-DMQL: a semantic trajectory data mining query language. Int Journal of Geographical Information Science 23:1245–1276

Bogorny V, Renso C, de Aquino AR, de Lucca Siqueira F, Alvares LO (2013) CONSTAnT—a conceptual data model for semantic trajectories of moving objects. Transactions of GIS 18(1):66–88

Cagnacci F, Boitani L, Powell RA, Boyce MS (2010) Challenges and opportunities of using GPS location data in animal ecology. Philosophical Transactions of the Royal Society of London: Biological Sciences 365(1550)

Damiani ML, Silvestri C, Bertino E (2011) Fine-grained cloaking of sensitive positions in location-sharing applications. IEEE Pervasive Computing 10(4):64–72

Guc B, May M, Saygin Y, Korner C (2008) Semantic annotation of GPS trajectories. In: Proceedings of AGILE

Krumm J, Horvitz E (2006) Predestination: inferring destinations from partial trajectories. In: Proceedings of UbiComp

Liao L, Fox D, Kautz H (2005) Location-based activity recognition using relational markov networks. In: Proceedings of IJCAI

Monreale A, Trasarti R, Pedreschi D, Renso C, Bogorny V (2011) C-safety: a framework for the anonymization of semantic trajectories. Transactions on Data Privacy 4(2):73–101

Palma AT, Bogorny V, Kuijpers B, Alvares LO (2008) A clustering-based approach for discovering interesting places in trajectories. In: Proceedings of ACM-SAC

Parent C, Spaccapietra S, Renso C, Andrienko G, Andrienko N, Bogorny V, Damiani ML, Gkoulalas-Divanis A, Macedo JA, Pelekis N, Theodoridis Y, Yan Z (2013) Semantic trajectories modeling and analysis. ACM Comput Surv 45(4):Article no. 42

Renso C, Baglioni M, de Macedo JAF, Trasarti R, Wachowicz M (2012) How you move reveals who you are: Understanding human behavior by analyzing trajectory data. Knowledge and Information Systems 37(2):331–362

Rocha JAM, Times VC, Oliveira G, Alvares LO, Bogorny V (2010) DB-SMoT: a direction-based spatio-temporal clustering method. In: Proceedings of IS

Spaccapietra S, Parent C (2011) Adding meaning to your steps. In: Proceedings of ER

Spaccapietra S, Parent C, Damiani ML, Macedo JA, Porto F, Vangenot C (2008) A conceptual view on trajectories. Data and Knowledge Engineering 65(1):126–146

Spinsanti L, Celli F, Renso C (2010) Where you stop is who you are: understanding peoples' activities. In: Proceedings of BMI

Wu D, Cong G, Jensen CS (2012) A framework for efficient spatial web object retrieval. The VLDB Journal 21(6):797–822

Yan Z (2011) Semantic trajectories: computing and understanding mobility data. PhD thesis, EPFL. http://infoscience.epfl.ch/record/167178

Yan Z, Parent C, Spaccapietra S, Chakraborty D (2010) A hybrid model and computing platform for spatio-semantic trajectories. In: Proceedings of ESWC

Yan Z, Chakraborty D, Parent C, Spaccapietra S, Aberer K (2011a) SeMiTri: a framework for semantic annotation of heterogeneous trajectories. In: Proceedings of EDBT

Yan Z, Giatrakos N, Katsikaros V, Pelekis N, Theodoridis Y (2011b) SeTraStream: semantic-aware trajectory construction over streaming movement data. In: Proceedings of SSTD

Yan Z, Chakraborty D, Parent C, Spaccapietra S, Aberer K (2012) Semantic trajectories: mobility data computation and annotation. ACM Trans Intell Syst Technol 9(4):Article no. 49

Yigitoglu E, Damiani M L, Abul O, Silvestri C (2012) Privacy-preserving sharing of sensitive semantic locations under road-network constraints. In: Proceedings of MDM

Zheng Y, Chen Y, Xie X, Ma WY (2010) Understanding transportation modes based on GPS data for Web applications. ACM Trans Web 4(1):Article no. 1

Zimmermann M, Kirste T, Spiliopoulou M (2009) Finding stops in error-prone trajectories of moving objects with time-based clustering. In: Proceedings of IMC

Chapter 10
The Case of Big Mobility Data

Trillions of bytes of information are being collected by companies about their customers, suppliers, and operations. Devices, such as mobile phones, tablets, smart energy meters, automobiles, industrial machines, carry millions of networked sensors which create huge portions of data that are channeled in the Internet of Things (IoT). In addition, the massive participation of individuals on social networking sites will continue to fuel exponential data growth. Hence, the enormous amount of data being produced in our world during the last decades has posed new challenges in the world of data management. Various methods and technologies have been developed and adapted in order to store, manage, analyze, and visualize the complex vast quantities of data in an efficient way. In this chapter, we introduce the reader to the concept of Big Data management and, then, we highlight interesting works made so far on the spatial and mobility domains. The area is still in its infancy and a strong wave of research results and, as a consequence, commercial products is expected to come up the forthcoming years.

10.1 Introduction to Big Data

The term "Big Data" refers to a collection of (usually) unstructured datasets, so large, diverse, and complex that is beyond the ability of typical database management tools to handle. The new challenges posed to the data management community include capturing, storing, indexing, searching, analyzing, and visualizing this kind of data. Nowadays, the volume of big data may vary from a few terabytes to several petabytes, depending on the application domain. The growth of the size of the datasets over time occurs due to the increasing use of cameras, microphones, radiofrequency identification readers, aerial sensory technologies, information-sensing mobile devices, software logs, and wireless sensors networks. Apart from the spread of sensor-based technologies, the trend to larger and larger datasets is also due to the combination of already large sets of related data, aiming e.g. at preventing diseases,

N. Pelekis and Y. Theodoridis, *Mobility Data Management and Exploration*,
DOI 10.1007/978-1-4939-0392-4_10, © Springer Science+Business Media New York 2014

discovering trends in the business world, determining real-time roadway traffic conditions, etc.

It is obvious that, handling big data is a difficult (if not impossible) task when using centralized database systems and desktop visualization and statistics packages, demanding an alternative approach that massively uses parallel software, running on numerous (in the order of thousands) servers. To gain value from big data, organizations tend to utilize new technologies (e.g. storage and manipulation software) and new analytical tools. Inheritance systems and incompatible standards and formats often prevent the integration of data and sophisticated analytics that create value from big data. Newly raised problems and the increasing computing power drive the deployment of new analytical techniques. For this purpose, there exists a growing number of technologies to analyze, store, manage, and manipulate big data, including Data Warehouses (DW), Business Intelligence (BI) tools, Cloud computing, etc.

As far as it concerns storing and managing big data there exist two traditional paradigms: parallel and distributed databases. The common field of both is that they are decentralized, meaning that they try to distribute data and/or computing power to multiple nodes over a network. In particular, a parallel database system pursues to increase performance through parallelization of various tasks, such as building indexes, loading data, evaluating and executing queries. Parallel databases may follow shared-disk, shared-memory or shared-nothing architecture, as illustrated in Fig. 10.1. In particular:

- the architecture of *shared-memory* systems consists of multiple CPUs and a shared memory, with the access of each CPU to the shared memory taking place over a common bus;
- in *shared-disk* systems, every node has one or more CPUs and its related memory (nodes do not share memory), while the communication takes place above a mutual high-speed bus, and every node has access to the shared disks;
- finally, in *shared-nothing* systems, each node is independent and self-sufficient in terms of memory and disk storage.

Depending on the adopted architecture, data may be stored in a distributed fashion or not and try to improve processing and input/output tasks by using multiple CPUs and disks in parallel. Moreover, parallel processing is divided in *pipeline parallelism*, where numerous computers perform one step each in a multi-step process, and *partition parallelism*, where numerous computers perform the same operation over different fragments of data.

On the other hand, a distributed database system is a collection of logically related databases that co-operate in a transparent way. The term 'transparent' implies that users may access data irrespective of the database it is physically stored, as if all data were stored in a single database locally. This implies "location independence", i.e. the user is not aware of the location of the data, while it is possible for the data to move from one physical location to another transparently to the user. Distributed databases are usually deployed in a shared-nothing architecture, which leads to reduced communication overhead and better performance. Furthermore, the hazard of a failure on a site is eliminated, i.e. if a server fails, then the only part of

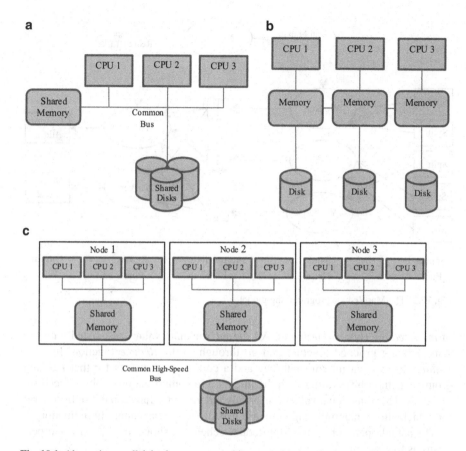

Fig. 10.1 Alternative parallel database system architectures: (**a**) shared-memory; (**b**) shared-nothing; (**c**) shared-disk

the system that is affected is the corresponding local site and the rest of the system remains functional and available.

As one could easily observe, there is overlap between parallel and distributed databases. In fact, nowadays, the predominant technologies, which can cope with big data, are mostly classified as "shared nothing". Cases of such success stories include Google File System, MapReduce, Hadoop, Big Table, HBase, Cassandra, etc.

10.2 The MapReduce Programming Model

MapReduce is a model of distributed programming created in order to process a particularly large volume of data, which are stored in different sites. This model is applicable to a wide range of use cases and is organized in a two-step Map and Reduce process. *Map* takes as input a set of (*key, value*) pairs and produces a group

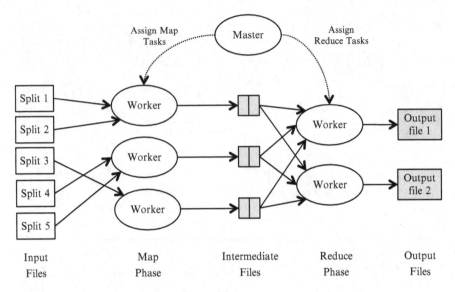

Fig. 10.2 The MapReduce programming model

of *intermediate* (*key, value*) pairs. All the intermediate values associated with the same key are grouped together and go through to the *Reduce* function. In turn, *Reduce* receives an intermediate key and a collection of values for that key and groups together these values with the intention to compose a probably reduced set of values. The transitional values are usually fed to reduce function through an iterator, which allows manipulating immense lists of values that cannot fit in memory.

The actual operation of the MapReduce scheme is shown in Fig. 10.2 and performs as follows:

1. The input files are split into M pieces and several copies of the program are initialized on a set of machines which belong on a cluster.
2. One of these machines is set to be the *master*, while the rest are set to be *workers*. The master assigns either *Map* or *Reduce* tasks to the idle workers. There are M map and R reduce operations, and each worker is assigned a *Map* or *Reduce* task.
3. Each worker, which has been assigned a *Map* task, receives as input the corresponding input split and extracts from it the (*key, value*) pairs. Subsequently, these pairs are delivered to the *Map* function where the intermediate (*key, value*) pairs are emitted and buffered in the memory.
4. The local disk is divided into R partitions, which are defined by the partitioning function, and at regular time intervals some of the intermediate (*key, value*) pairs are written to it. The position on the disk, that each pair is written, is transmitted to the master node, thus making it responsible for passing these locations to workers that have been assigned a *Reduce* task.
5. As long as a *Reduce* worker becomes aware of the positions of the intermediate (*key, value*) pairs, connects to the corresponding *Map* worker through remote procedure calls and reads the data. After the *Reduce* worker has read all of the data, then it sorts them by the key. In this way, the same keys are grouped together.

6. Subsequently, the *Reduce* worker goes through the ordered intermediate (*key*, *value*) pairs and for every distinct key the respective (*key*, *value*) pair is fed to the *Reduce* function, while the outcome is appended to an output file for this reduce partition.
7. When all the *Map* and *Reduce* tasks have been performed, the master returns the control to the user program.

As already mentioned, the master keeps track of the *Map* and *Reduce* tasks by storing their state (could be 'idle' or 'in-progress' or 'complete') and the identity of the worker machine that has undertaken it. A machine failure control mechanism is also active during the entire process. As far as it concerns a worker failure, the master pings every worker at regular intervals. If the worker does not respond within a time period then the master considers the worker as failed and resets the status of every map task, that was assigned to it, to idle, and it is transferred to another worker. If a map task is completed but the machine that it is stored in fails, then it has to be re-executed, unlike completed reduce tasks, that don't have to be re-executed, because the output of *Reduce* tasks is stored in a global file system. In general, MapReduce is durable in large-scale worker failures, a fact corresponding to one of the basic assumptions behind MapReduce, namely that a failure is sure to occur. On the other hand, MapReduce assumes that a master failure is unlikely to happen due to the fact that there is only one master node. So if the master fails all computations are aborted. However, periodically the master writes checkpoints of the data structures mentioned above and when the master fails a new copy can be started from the last checkpoint.

Another important issue is the size of M and R. In an ideal case, M and R should be much larger than the cardinality of the workers. By getting every worker to perform many different tasks the dynamic load balancing is improved and, furthermore, in case of a worker failure the recovery is much faster.

10.2.1 Hadoop

Hadoop is the prevalent implementation, of the MapReduce paradigm. It provides a distributed file system, which is designed for storing extremely large datasets and streaming them very efficiently to user applications. Since Hadoop is based on MapReduce, its most important feature is the partitioning of both data and computation to several machines. Hadoop, as shown in Fig. 10.3, follows a master/slave architecture. The Master Node consists of a *JobTracker*, a *NameNode*, and a Secondary *NameNode*. Each Slave Node consists of a *TaskTracker* and a *DataNode*.
More specifically:

– *JobTracker*: is the component where the user application meets Hadoop. After loading the application on cluster, the *JobTracker* determines the execution plan defining which files it will process, assigns tasks to different nodes and supervises all tasks in progress. If a job fails, the *JobTracker* will automatically re-launch the job, possibly on a different node, for a predefined limit of retries.

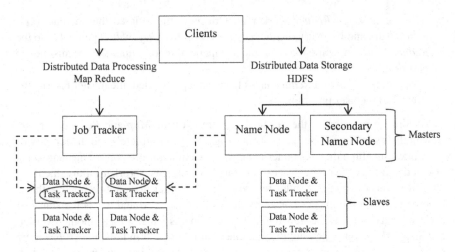

Fig. 10.3 Hadoop architecture

It also provides the user interface with a MapReduce cluster. There is only one *JobTracker* service per Hadoop cluster. It typically runs on a server, such as the master node of a cluster. In correspondence with MapReduce, the *JobTracker* plays the role of the master concerning only the scheduling and supervision of the *Map* and *Reduce* tasks.

- *NameNode*: The Hadoop Distributed File System (HDFS) stores application data and file system metadata separately. Application data are kept on a set of servers called *DataNodes*, while the file system metadata are stored on a separate dedicated server called *NameNode*. Each Hadoop *cluster* consists of a single *NameNode*, which contains a hierarchy of the files and directories that are represented by *inodes*. These *inodes* keep track of characteristics such as permissions, namespace and disk space quotas, modification and access times. In addition, it preserves the namespace tree and the physical location of the file blocks on the *DataNodes*.

- *TaskTracker*: Slave Nodes make up the vast majority of machines and perform all the *dirty* work of storing data and execution. Each slave runs a *DataNode* and a *TaskTracker* job. *TaskTracker* is a perpetual (daemon) process that receives instructions from master nodes. Actually, the *TaskTracker* is a slave to the *JobTracker* and is responsible for the execution of a specific job on every slave node.

- *DataNodes*: Each data block copy is delineated by two files in the local drive of the *DataNode* that hosts it. The first holds the actual data while the second contains some metadata, such as checksums, of the data block as well as the block's generation stamp.

Hadoop is installed on client machines, the functionality of which is divided into MapReduce clients and HDFS clients. The role of MapReduce clients is to request MapReduce jobs describing how data processing will be done and then recovering the results from these jobs. The HDFS clients are dealing with loading and reading data from the cluster. In particular, when an HDFS client wants to read a file, it contacts the *NameNode* requiring the positions of the data blocks that constitute the desired file, and then reads these blocks straight from the DataNodes closest to the client. In the case of loading data, the client asks the *NameNode* to suggest a group of three *DataNodes* to receive the block replicas. Then, the client writes data to these *DataNodes* in a pipeline style. Due to the fact that each *DataNode* can perform several application tasks simultaneously, Hadoop can hold up to tens of thousands of HDFS clients per cluster.

Current Hadoop's implementation of MapReduce does not support any kind of indexing mechanism. This is not to be viewed as a drawback of MapReduce; actually, it is something that MapReduce has not been designed for. MapReduce has been designed for one-time processing of large data sets in batch mode. In the future, empowering MapReduce with indexing mechanisms will make it suitable for real time data analysis.

10.2.2   HadoopDB

HadoopDB appears to be a very promising implementation that joins parallel database systems and the MapReduce paradigm (more specifically, the Hadoop implementation of MapReduce). The idea behind HadoopDB is to connect a set of single-node database systems by employing Hadoop as the network communication layer and task coordinator. Queries are basically parallelized across the nodes via the MapReduce framework; nevertheless, HadoopDB tries to push as much of the single node query work as possible inside the corresponding node databases. The main achievements of HadoopDB are heterogeneity and fault tolerance, by adopting the corresponding features from Hadoop, however it reaches the performance of parallel databases by performing as much of the query processing as possible inside the database engine.

HadoopDB architecture, illustrated in Fig. 10.4, extends Hadoop's characteristics providing the following four components:

- **Database Connector**. It is the means for communication between JobTrackers and independent database systems residing on nodes in the cluster. Each MapReduce job supplies the Connector with an SQL query and the connection parameters (such as which JDBC driver to use, query fetch size and other query parameters). The Connector connects to the database, executes the query, and returns results as (*key, value*) pairs. As compared to the Hadoop framework, the databases are data sources similar to data blocks in HDFS.
- **Data Loader**. It manages the parallel load of data from a file on the databases systems.
- **Catalog**. The catalog maintains meta-information about the databases and is crucial for the creation of the query. It includes connection parameters and metadata, such data partitioning properties and replica locations.
- **SQL MapReduce planner**. Provides an SQL interface on HadoopDB.

One of the drawbacks of HadoopDB that makes it unsuitable in the realm of large-scale data processing is the lack of fault tolerance at the data layer. This is due to the fact that partitioning of the raw data and uploading them to individual

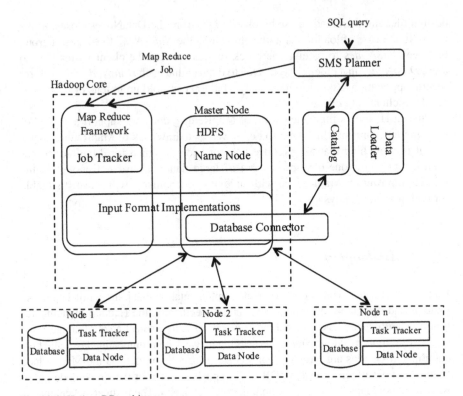

Fig. 10.4 HadoopDB architecture

database nodes is no more supervised by the Hadoop framework. While HadoopDB integrates the power of efficient DBMS technology with MapReduce, yet it seems impractical to carry out large scale data processing.

10.3   Handling Big Spatial Data

An interesting instantiation of the big data management problem is the manipulation of spatial data. Spatial data consist of numerous longitude-latitude pairs representing with extreme accuracy the geometry of spatial objects, such as polygons (e.g. towns), lines (e.g. rivers), points (e.g. landmarks) etc. Moreover, in the Big Data era there exist several real world applications, which produce vast quantities of spatial data, such as the Earth Observing System (EOS) by NASA that generates up to 1TB of data on a daily basis. At the same time spatial data operations are computationally and memory intensive. A typical solution to deal with this issue is to distribute the spatial operations amongst several computational nodes. Parallel spatial databases try to achieve this but at a very small scale (i.e. few 10's of nodes). Another approach would be to use distributed approaches, such as MapReduce, since spatial data is clearly distributable by exploiting on the spatial locality of the objects. However, by adopting this approach, one would have to sacrifice any indexing mechanism and operate on unstructured data. Nevertheless, in the literature there has been proposed a method for building R-tree spatial indexes by utilizing the MapReduce framework. Finally, there is an approach that tries to incorporate the advantages of parallel spatial DBMS and the MapReduce paradigm in a single architecture.

10.3.1 MapReduce-Based Approaches

An interesting approach that utilizes the MapReduce paradigm in order to store, query, and index big spatial data adopts a data partitioning strategy based on the spatial dimension of data in order to attain a high I/O throughput. More specifically, the partitioning mechanism divides the data into blocks by taking into account the geographic space and a block size threshold. These blocks are equally dispersed on the cluster nodes. As a result, the objects that hold adjacent spatial information are sequentially stored by utilizing a space-filling curve in order to retain the geographic proximity of the objects. In practice, most of the clients that make use of a spatial application concentrate on a relatively small region and hence the queries that they pose regard spatial objects within that area. By taking this into account, it is obvious that clients can be served simultaneously and in parallel by different nodes, and nearby objects can be streamed to clients in sequence without random I/O seeks.

The second step aims to increase data retrieval efficiency by designing a two-level distributed spatial index for pruning efficiently the search space. At the first level, there exists the global index which makes use of a Quadtree-based indexing mechanism and is used in order to find data blocks. The second level consists of a Hilbert-ordering local index, which is used to locate spatial objects. At the third step, an "indexing + MapReduce" data processing architecture is adopted, aiming to increase the efficiency of spatial queries. This architecture employs parallel processing techniques to deliver both inter-query and intra-query parallelism, and thus can lead to reducing the execution time of individual spatial queries and at the same time afford a great number of simultaneous spatial queries.

Another interesting approach for tackling the problem of big spatial data using MapReduce aims at constructing an R-tree indexing structure in a parallel and distributed environment. Let us assume that each object has a unique identifier, $o.id$ and some spatial properties, $o.P$. Other attributes are also possible to exist but these two are sufficient enough for the R-tree construction. The spatial information of each object is approximated by its MBR, employed for the construction of R-tree nodes. Furthermore, the object identifiers are used for referencing the objects stored in the R-tree leaf nodes. The R-tree construction method consists of three phases, as illustrated in Fig. 10.5.

The first phase partitions the spatial objects into groups, at the second phase, a small R-tree is created for each group, and at the third phase, these small R-trees are merged into the final R-tree. The MapReduce framework is utilized in the first two phases, while the last phase is executed sequentially outside the cluster because it is not computationally demanding.

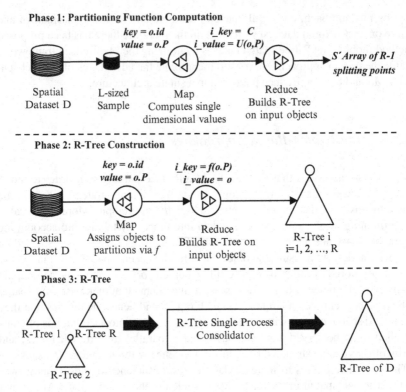

Fig. 10.5 Phases involved in building an R-Tree index for a dataset in MapReduce

In more detail, the three phases operate as follows:

- The first phase takes as input the spatial data set D and the number R of partitions, in which the data set will be partitioned. Then, the goal is to infer a partitioning function f, so as each object of the data set is assigned to one of the partitions. This partitioning function is computed in such way so as when f is applied in D, ideally, R equally sized partitions are produced. The output of the first phase is the function f (note that the actual partitioning does not take place in this phase).
- The input of the second phase is again the spatial dataset D and the partitioning function f resulted at the first phase. At this phase, D is split into R partitions according to f. Afterwards, a number of R small and independent R-trees are built in parallel by the respective Reducers. So, the output of this phase is a forest of R independent R-trees.
- The final phase takes as input the forest of R independent R-trees and tries to combine them into a unified R-tree index of the data set under a single root node. As already mentioned, this phase is not computationally intensive and it is executed outside the cluster.

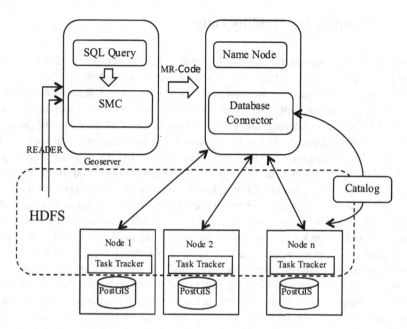

Fig. 10.6 Overall architecture of a hybrid PostGIS—HadoopDB implementation. *Source*: Sagar (2012)

10.3.2   A Hybrid Spatial DBMS—MapReduce Approach

Recently, there are plenty of studies aiming to formulate spatial operations as a MapReduce problem. However, the MapReduce platform has originally been designed for a single procedure of data processing. In contrast, the majority of spatial applications do not usually follow the model of single processing, as continuous monitoring and progressive analysis is required to derive interesting and actionable knowledge. In this light, it turns out that MapReduce is not the most suitable programming model for conducting spatial analysis. Also, the nonexistence of indexing mechanisms strengthens this argument. On the other hand, spatial DBMS are very widespread, mature and, in contrast to MapReduce, optimized for spatial data management.

Towards the objective to couple the two paradigms, a hybrid spatial DBMS—MapReduce architecture, illustrated in Fig. 10.6, is built as follows. The HadoopDB platform constitutes the basis on which the rest of the modules are placed. At the second level, the storage system of HadoopDB is replaced by PostGIS database servers, which store the spatial information. At the third level, a front-end tool (such as GeoServer) provides high-level SQL processing environment for spatial data to the end user.

10.4   Handling Big Mobility Data

Having already examined the way that big spatial data are being treated in the era of "Big Data" management, it becomes obvious that the necessity of manipulating big mobility data (e.g. trajectories or current locations of moving objects) is even larger due to the immense amount of data being produced on a daily basis and the increasing need for storing, querying, analyzing, and extracting knowledge out of them.

Mobility data applications can be classified into two categories: historical and real-time. The management of historical data implies that the data are stored in a single or multiple sites in an archival mode. Historical data are used for offline querying, analysis, and extraction of useful knowledge. As such, the respective applications try to optimize querying over the entire dataset. An optimization process could involve indexing mechanisms, storing the data in a distributed way, etc. Usually, this process is application specific, meaning that depending on the queries that an application desires to apply over the data (e.g. range, NN queries, etc.), it tunes its indexing mechanisms and storage infrastructure in order to efficiently answer the queries. On the contrary, real-time mobility data get updated continuously (or in batch mode) with the current location of the objects. At the same time, the past locations of each object should be maintained for querying and analysis purposes. Based on the assumption that past locations of the objects are less likely to be retrieved, the effort is being focused on managing the recent and current versions of the data in such a way that they can be efficiently updated and queried.

Big mobility data management deals with amounts of data so huge that the traditional database systems are unable to handle. For this reason, the approaches for handling this kind of data involve non-traditional, distributed and parallel systems. In the literature, there are few recent efforts that utilize the MapReduce framework, which either deal with historical mobility data, or they are hybrid in the sense that they also try to tackle real-time requirements.

10.4.1   Offline Mobility Data Analytics

Regarding offline analysis on big mobility data an interesting approach is called CloST, a scalable spatiotemporal data storage system that supports data analytics using Hadoop. It proposes a way for storing big spatiotemporal data using the Map-Reduce framework, so as to optimize two types of queries, *single-object spatiotemporal range queries*, $Q(I, S, T)$, versus *all-object spatiotemporal range queries*, $Q(S, T)$, where S is the spatial range, T is the temporal range and I is the object identifier.

CloST stores spatiotemporal data in tables, in the form of $(Oid, Loc, Time, A_1, \ldots, A_n)$. The first three attributes are the core attributes, where Oid is the object identifier, Loc is a spatial point, and $Time$ is a timestamp. To enable the efficient processing of both types of queries, a three-level hierarchical partitioning is employed, as illustrated in Fig. 10.7. At the first level, the data are divided into a number of partitions according to hash values of the object ids and coarse ranges of time. At the

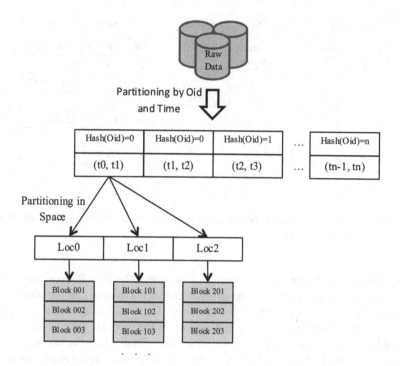

Fig. 10.7 Hierarchical partitioning in CloST

second level, each partition derived from the first level is divided into a set of partitions according to a spatial index on the location attribute. Finally, at the third level, the actual data are found. Here, levels 1 and 2 serve as indexes for level 3.

In order to achieve its goal, given the fact that Hadoop is employed, CloST devises a space efficient file format to actually store data in HDFS. Records are grouped by object id and each group is stored in a file section. An in-block index is placed at the beginning of each file in order to map each object into the corresponding position inside the file section. Then, for each section, records are sorted by time and then organized in a column-store fashion. To improve the storage efficiency, CloST compresses data first at column-level where numeric values are encoded by either delta encoding or running-length encoding. Afterwards, data are compressed at section-level where all data inside a section are compressed once more using gzip to further decrease the data size.

Furthermore, a Metadata table exists, containing information about mappings from buckets (level 1) to regions (level 2) and from regions to blocks (level3). This table is stored together with the data, under the same directory, which makes it easy to automatically replicate and distribute it along with the actual data. Finally, for a single object query, the Metadata table is used together with the in-block indexes in

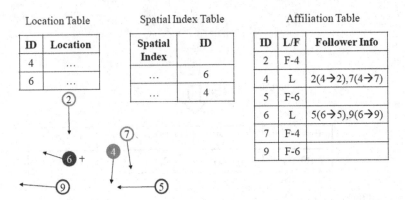

Location Table   Spatial Index Table   Affiliation Table

Fig. 10.8 The content of Location, Spatial Index and Affiliation tables in MOIST for six objects organized in two object schools

order to locate the desired data. For an all-object query, the Map-Reduce framework is employed in order to execute the query in parallel. Primarily, all the blocks that intersect the spatiotemporal range are retrieved. Then, each Map task sequentially scans the corresponding block file and assembles the records. Then, it filters each record by S and T in order to output the final result. Finally, in order to optimize the blocks size, CloST retains a query log of the recently posed queries and periodically tunes both spatial and temporal partitioning.

10.4.2   Hybrid Historical—Real-Time Approaches Using MapReduce

Regarding mobility-aware services (e.g. LBS) that necessitate the vast execution of update operations in order to keep the data up-to-date, there has been recently proposed an approach based on the MapReduce paradigm. The so-called MOIST (Moving Object Indexer with School Tracking) aims to reduce the update latency and at the same time to efficiently process historic information.

As illustrated in Fig. 10.8, the basic structures of MOIST are the Location Table the Spatial Index Table and the Affiliation Table. The Location Table stores the location and velocity information for each object, in such a way that queries on an individual object can be answered efficiently. MOIST clusters data into object schools (OS), consisting of objects that have spatial proximity and similar velocities. In this way, redundant updates can be reduced and consequently the update latency can be limited. The Affiliation Table is used to keep track of OS, by keeping information about the mapping for each cluster and is keyed by the behavior (i.e. the leader's speed) of that cluster. In the Spatial Index Table the leader object ID is stored as the value associated with the key of a cell (i.e. Spatial Index), after the partitioning of the space with a regular grid, if the object resides inside that cell.

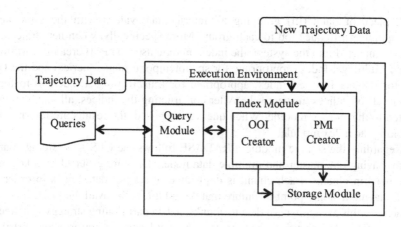

Fig. 10.9 PRADASE execution overview and framework

The data are indexed with Google's S2Cell indexer. The space is divided into a number of cells of different resolutions and a space filling curve (Hilbert Curve) is constructed by linking all cells in a space by the sequential order of their Spatial Indexes (derived from the Hilbert values of the cells). MOIST groups objects nearby to each other and with similar moving behavior into one school. This is based on the observation that objects close to each other often move with similar speed and trajectories. In order to keep track of an OS, MOIST keeps track of the leader object only, and records the distance between the follower objects and the leader. OS are preserved and transformed when an update arrives. An object leaves its OS when its distance to its leader surpasses a given threshold. After an object departs an OS, it becomes a leader of a new OS. OS are merged periodically using a fast clustering method.

Finally, aged data are flushed onto disk so that the history of the moving objects can be analyzed. A naïve approach for achieving this is to move an updated object location onto the disk before a new update arrives. This can lead to a large number of disk IO's and, consequently, to a latency penalty. To reduce this latency overhead, a double-buffering (or Ping-Pong buffering) scheme can be utilized. While updates are occurring on one memory buffer, another one is moved onto the disk. What should be ensured in this scheme is that the time required for flushing aged data from the buffer onto the disk is less than the time required to fill the other buffer in memory.

Another hybrid approach, where not only historical data but also updates of moving objects are treated by the MapReduce framework, is called PRADASE (Query Processing of Massive Trajectory Data based on MapReduce). It is based on a GFS-style storage which supports append only mode. The master node is in charge for key to (data) nodes mapping and lookup (via hashing). So, if the key is given, the corresponding data can be returned automatically. Here, the queries under optimization are spatiotemporal range and trajectory based queries.

As shown in Fig. 10.9, the framework consists of three modules: the storage, the index, and the query module, respectively. The execution process is performed in four steps: (i) splitting input data into several pieces; (ii) executing a Map job for each piece of data; (iii) grouping all intermediate values with the same key; (iv) applying a Reduce job to each group. More specifically, when new trajectory data is imported into the system, the index module uses the PMI creator (Partition-based Multilevel Index, appropriate for spatiotemporal range queries) and the OII creator (Object Inverted Index, appropriate for trajectory-based queries) to form two kinds of indices, respectively. After constructing the indices, all data is transmitted to the storage module. When queries are posed, the results are returned by invoking the index module.

Regarding the storage module, PRADASE follows the GFS-style storage paradigm, having one master and multiple data nodes. Data are grouped by a key and organized in chunks. Each chunk is duplicated in several data nodes in order to avoid having a single point of failure that would affect the availability of the data. To achieve this assignment of data to chunks, a data partitioning strategy is defined. An acceptable data partitioning strategy is one that leads to a uniform data distribution, thus maximizing the efficiency of the whole cluster and avoiding the occurrence of a load imbalance. In order to guarantee a better load balance on highly skewed trajectory data, a hybrid method, which employs a different static partitioning strategy for each time period, is proposed. To achieve this, a monitoring mechanism is employed, which records the size of all chunks. In case it detects an imbalance in the size of the data of each chunk, a new partitioning strategy is trained. This new strategy affects only the new updated data, leaving the data which has already been imported into the system intact. This means that, regarding the whole time dimension the partitioning is dynamic, but within each time period is static. Having partitioned the (temporal or spatiotemporal) space, each trajectory is assigned to at least one partition based on its spatiotemporal properties. A line segment is assigned to one partition if and only if it is fully covered by one partition. If a line segment extends to more than one partitions, then it needs to be divided into several line segments, where each line segment is fully covered by one partition.

Regarding the index module, two different indices are used (PMI and OII, respectively) in order to optimize the two types of queries of interest. In more detail, by taking into account the requirement that each chunk contains trajectory segments that belong to the same partition, a hash index is built in order to speed up range queries. So, for any given spatiotemporal range, all candidate partitions, where the given query range covers, are retrieved by invoking the space partition strategy. The results of the query will be contained in these candidate partitions. Hence, only a few candidate chunks need to be scanned instead of all data chunks. Furthermore, in each chunk, a multi-level index is built. In detail, in order to accelerate trajectory based queries, an OII index for each moving object is employed. For each object, all historical trajectories are collected and stored together. The data model of OII is presented by a triple $<OID, Pl, T>$, where OID is the object identifier, Pl is the object's current location (including spatial location and timestamp), and T is the object's historical trajectory (set of line segments). Hence, given any OID, the last known location and all corresponding trajectories will be returned quickly. For object-oriented queries, a two-level hash index is designed where at the first level, the data node where the data chunk is stored, is returned, and at the second level, trajectory data are addressed.

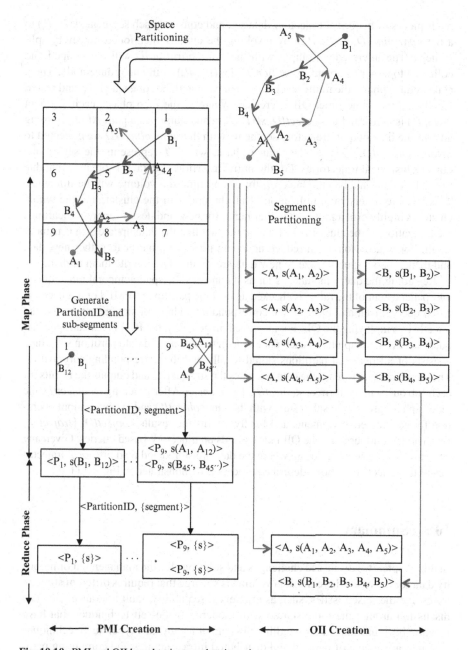

Fig. 10.10 PMI and OII based trajectory data insertion

In order to facilitate massive trajectory dataset insertion, a MapReduce-like model is utilized (shown in Fig. 10.10). When the data insertion starts, two MapReduce sub-processes are triggered, the process of PMI creation and the process of OII creation. Concerning the PMI creation, at the map phase each data node reads pieces of historical trajectory data set, and converts each segment $sk(P_i, P_j)$ to a list $<partitionID, sk(P_i, P_j)>$ by invoking the already mentioned trajectory split strategy. The intermediate output with the same *partitionID* will be grouped and collected together to obtain $<partitionID, \{s_1 (P_m, P_n)\}>$ in one reduce node. Then, at the reduce phase, the multi-level index can be generated for each group and stored locally. At the same time, OII is created, where at the map phase, each segment $sk(P_i, P_j)$ is converted to $<objectID, sk(P_i, P_j)>$ and output as a pair. After grouping all intermediate output, trajectory segments with the same *objectID* are collected to obtain $<objectID, \{T\}>$ at the reduce phase, where T represents the set of one object's historical trajectories. For the initial insertion of data, it is obvious that due to the fact that data has mounted up for a long time the volume will be immense. This leads to a very high utilization of all the nodes in the cluster, and the whole cluster is highly efficient. On the other hand, for new updates of reported locations, the utilization of the cluster is low due to the fact that the data update size is usually small. For that reason, each reduce node can preserve a buffer to gather new data and then, when a specific size of new data are accumulated, write them to disk.

Regarding the query module, as far as it concerns the spatiotemporal range query, the spatial extent of the range is checked against the partition of PMI. For a trajectory-based query, given an object id, the corresponding address of storage can be located directly by employing the OII. For any given range query, the PMI index can speed up querying. At the map phase, each data node generates a candidate partition set which consists of all potential partitions stored locally and the corresponding sub queries. Then, each data node reads the data, executes the sub queries and outputs the results for each sub query in the form of $<objectID, \{segment\}>$. After grouping by *objectID*, the reduce phase groups the sub results with the same *objectID* and executes a join operation for each object's segments and finally returns the results $<objectID, \{trajectory segments\}>$. Furthermore, the OII index speeds up trajectory-based queries. Given any object ID, the system can locate where the data are stored and read historical trajectories of the object. The above-described process is demonstrated in Fig. 10.11.

10.5   Summary

In this chapter, we discussed challenges and solutions for the management of mobility data for the case where the data volume is so large that requires different strategy to resolve the raised issues, such as efficiency, scalability, fault tolerance, etc. The discussion around the topic started with modern off-the-self techniques that have recently been introduced for handling big data of legacy data type. Such techniques take into advantage of parallel and distributed infrastructures and computing paradigms, such as the MapReduce programming model. The proposed methods that focus on mobility data, for the moment, propose minor improvements with respect to traditional data and one expects that more radical steps will follow very soon, due to the emerging interest of the community. Similarly, more advanced and intricate analytic operations, such as mining tasks, have already started to be investigated. This is anticipated to be the next wave of development in the field.

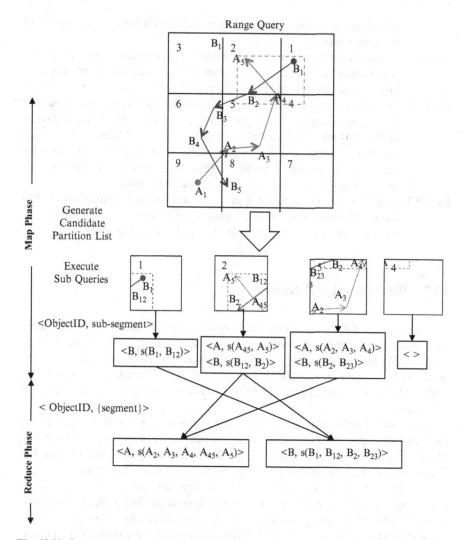

Fig. 10.11 Spatiotemporal range query in PRADASE

10.6 Exercises

Ex. 10.1. What are the advantages and disadvantages of the original Hadoop with respect to the HadoopDB framework?

Ex. 10.2. Search in the literature for various alternative data management models, such as NoSQL databases, cloud-based solutions, graph data management systems, column stores. What are their differences? Which of them are directly applicable to the mobility data case? If not directly applicable, think of the adaptations they would be required, either to these approaches or to the representation model of mobility data, in order to be working solutions.

Ex. 10.3. Devise an algorithm for the R-tree consolidator presented in Sect. 10.3.1, which takes as input a forest of independent R-trees and combines them into a unified R-tree, under a single node.

Ex. 10.4. Consider as an example a couple of semantic trajectories (e.g. those that correspond to typical daily agendas of yours). Try to think of adaptations of the PMI and OII indexing approaches of the PRADASE framework that could capture the new requirements. Re-design Fig. 10.10 for the case of your semantic trajectories. How would you handle the semantic annotations that correspond to the textual information?

Ex. 10.5. Search in the literature for approaches that propose adaptations of data mining algorithms under the MapReduce programming model. Which of them are (or could be) applicable to spatial or mobility data?

10.7   Bibliographical Notes

Regarding parallel database management systems, DeWitt and Gray (1992) offers an insightful and detailed overview of those systems from the beginning of their existence. As far as it concerns introductory concepts and techniques for big traditional data, Manyika et al. (2011), Lee et al. (2011), Dittrich and Quiané-Ruiz (2012), Shim (2012) and Zhang et al. (2012) provide extended up-to-date overviews of the area.

The Google File System was introduced in Ghemawat et al. (2003), upon which the MapReduce programming model was presented by Dean and Ghemawat (2004). The distributed storage system for structured data called BigTable was presented in Chang et al. (2006). Borthakur (2007) proposed the Hadoop file system, an open-source implementation of the Google File System that utilizes the MapReduce framework. A detailed and perceptive comparison of the MapReduce framework versus Parallel DBMSs was provided in Pavlo et al. (2009). An effort for the architectural hybrid of MapReduce and DBMS technologies for analytical workloads, hence combining the advantages of the MapReduce framework and Parallel DBMSs, was made in HadoopDB (Abouzeid et al. 2009).

Regarding big spatial data management, the MapReduce-based approaches presented in Sect. 10.3.1 were introduced by Cary et al. (2009) and Zhong et al. (2012). The hybrid PostGIS-HadoopDB approach presented in Sect. 10.3.2 was proposed in Sagar (2012). Concerning big mobility data management, an interesting preliminary effort in the field includes Hadjieleftheriou et al. (2004). Finally, concerning the architectures presented in Sect. 10.4, the MOIST architecture was proposed in Jiang et al. (2012), the CloST system was presented in Tan et al. (2012), and the PRADASE framework was introduced in Ma et al. (2009).

References

Abouzeid A, Bajda-Pawlikowski K, Abadi D, Silberschatz A, Rasin A (2009) HadoopDB: an architectural hybrid of MapReduce and DBMS technologies for analytical workloads. Proceedings VLDB Endowment 2(1):922–933

Borthakur D (2007) The Hadoop distributed file system: architecture and design. Hadoop Proj Website 11:21

Cary A, Sun Z, Hristidis V, Rishe N (2009) Experiences on processing spatial data with mapreduce. In: Proceedings of SSDBM

Chang F, Dean J, Ghemawat S, Hsieh WC, Wallach DA, Burrows M, Chandra T, Fikes A, Gruber RE (2006) Bigtable: a distributed storage system for structured data. In: Proceedings of OSDI

Dean J, Ghemawat S (2004) MapReduce: simplified data processing on large clusters. In: Proceedings of OSDI

DeWitt D, Gray J (1992) Parallel database systems: the future of high performance database systems. Commun ACM 35(6):85–98

Dittrich J, Quiané-Ruiz JA (2012) Efficient big data processing in Hadoop MapReduce. Proceedings VLDB Endowment 5(12):2014–2015

Ghemawat S, Gobioff H, Leung ST (2003) The Google file system. In: Proceedings of SOSP

Hadjieleftheriou M, Kriakov V, Tao Y, Kollios G, Delis A, Tsotras VJ (2004) Spatio-temporal data services in a shared-nothing environment. In: Proceedings of SSDBM

Jiang J, Bao H, Chang EY, Li Y (2012) MOIST: a scalable and parallel moving object indexer with school tracking. Proceedings VLDB Endowment 5(12):1838–1849

Lee KH, Lee YJ, Choi H, Chung YD, Moon B (2011) Parallel data processing with MapReduce: a survey. ACM SIGMOD Rec 40(4):11–20

Ma Q, Yang B, Qian W, Zhou A (2009) Query processing of massive trajectory data based on MapReduce. In: Proceedings of cloud data management workshop

Manyika J, Chui M, Brown B, Bughin J, Dobbs R, Roxburgh C, Byers AH (2011) Big data: the next frontier for innovation, competition, and productivity. McKinsey Global Institute, San Francisco

Pavlo A, Paulson E, Rasin A, Abadi DJ, DeWitt DJ, Madden S, Stonebraker M (2009) A comparison of approaches to large-scale data analysis. In: Proceedings of SIGMOD

Sagar A (2012) Large spatial data computation on shared-nothing spatial DBMS cluster via MapReduce. MSc thesis, IIT, Bombay. Available at http://www.gise.cse.iitb.ac.in

Shim K (2012) MapReduce algorithms for big data analysis. Proceedings VLDB Endowment 5(12):2016–2017

Tan H, Luo W, Ni LM (2012) CloST: a Hadoop-based storage system for big spatio-temporal data analytics. In: Proceedings of CIKM

Zhang C, Li F, Jestes J (2012) Efficient parallel kNN joins for large data in MapReduce. In: Proceedings of EDBT

Zhong Y, Han J, Zhang T, Li Z, Fang J, Chen G (2012) Towards parallel spatial query processing for big spatial data. In: Proceedings of IPDPSW

Part V
Epilogue, Hands-on

As you set out for Ithaca, hope the voyage is a long one, full of adventure, full of discovery...

Constantine Cavafy

Chapter 11
Epilogue

This chapter concludes our 'tour' in the Mobility Data Management and Exploration field. We presented a step-by-step methodology to understand, manage, and exploit mobility data, from modeling aspects to issues arising due to the Big Data era.

It was 10 years ago when a seminal research activity dedicated to spatio-temporal databases concluded as follows: "... *CHOROCHRONOS has opened many avenues for research in spatio-temporal databases, but it also left us with lots of challenging research problems awaiting solution ... As an epilogue to this book, we would like to challenge the reader by discussing three important application areas and the role spatio-temporal databases can play in these ... Mobile and Wireless Computing ... Data Warehousing and Mining ... the Semantic Web ...*". Since then, all three then-open topics have been extensively studied and nice results have appeared in the literature. Also in this book, the reader may find related material in Parts I, II, and III, respectively.

In the 10 years that followed, a big bang happened in our everyday lives: that of *online social networking services*; content-oriented web sites, such as LinkedIn and MySpace were launched in 2002, followed by general-purpose sites, such as Facebook in 2004, Twitter in 2006, and Google+ in 2011, to name but a few. Location content was out of scope in the beginning but soon appeared to be one of the hot contents of social networks, resulting in extra services of general-purpose sites (Google Latitude in 2009, Facebook Places in 2010) or even new services oriented to the "where" information (Gowalla in 2007, Waze in 2008, Foursquare in 2009). Having the wide spread of this phenomenon in our mind, what would be the challenges in mobility data management and exploration in the years ahead? Several ideas are around, below we highlight two of them:

- Challenge 1: *Using movement information to improve our lives* (hence, *infomobility*). For instance, how and what for humans are moving in urban areas? how and what for wildlife animals are moving in their places?
- Challenge 2: *Spreading movement information to publicize our lives* (hence, *mobile-aware social networks*). For instance, what if more than one billion users of social networks add content about their movement?

N. Pelekis and Y. Theodoridis, *Mobility Data Management and Exploration*,   235
DOI 10.1007/978-1-4939-0392-4_11, © Springer Science+Business Media New York 2014

Infomobility is defined as *"the use and distribution of dynamic and selected multi-modal information to users, both pre-trip and, more importantly, on-trip, in pursuit of attaining higher traffic and transport efficiency as well as higher quality levels in travel experience by the users"*. It includes developments in Intelligent Transport Systems (ITS), such as real-time traffic management and control on the road network, traffic demand management, electronic message signing, electronic toll collection, multi-modal journey planning, etc. In this line, advances in mobility data management (real-time systems, big data management) and exploration (trajectory-based traffic analysis, future movement prediction) are more than welcome.

On the other hand, adding mobility (e.g. waypoints) in our social network profiles, thus shifting from location- to mobile-aware social networks, is not only a matter of size; how large could this database be, in what detail/simplification level should a trajectory be stored in such database. More interestingly, it is the motion patterns that could be mined for this data. As an example, a research challenge could include discovering *interaction patterns*. Consider, for example, a group of friends moving in an open festival area (a group of kids moving in an entertainment park, a group of university students moving in a campus, etc.) and *posting* their movement in a social network. Their collective movement could be analyzed online in order to classify this movement as *leadership* (the majority follows the steps of a leader), *path following* (the group moves so due to external constraints, e.g. a path that had to be followed), *goal seeking* (the group moves so due to the existence of a goal to be reached), etc. Advances in (privacy-aware) mobility data mining are expected in this domain. As a second example, consider a large collection of users' trajectories and queries like searching for users who travel mostly by bus, for users who follow the pattern home-work-home almost daily, for users who make similar stops with their friends in touristic areas, etc. Brand new semantic-aware mobile social network services could be built under this framework and, of course, clear privacy concerns also arise asking for solutions.

The above lines make clear that the next wave of scientific results in the area should be interdisciplinary: computer scientists, transportation engineers, geographers, and social scientists could smoothly cooperate, contributing with their own tools and perspective.

11.1 Bibliographical Notes

The quote on CHOROCHRONOS research activity is from Koubarakis et al. (2003). The definition of Infomobility appears in Ambrosino et al. (2010). Semantic-aware mobile social network services are discussed in Pelekis et al. (2013). Privacy issues that arise in modern location-aware social networking applications are discussed in Ruiz Vicente et al. (2011).

References

Ambrosino G, Boero M, Nelson JD, Romanazzo M (2010) Infomobility systems and sustainable transport service. ENEA, Rome

Koubarakis M, Theodoridis Y, Sellis TK (2003) Spatio-temporal databases in the years ahead. In: Sellis TK et al (eds) Spatio-temporal databases—the CHOROCHRONOS approach. Springer, Berlin, pp 346–347

Pelekis N, Theodoridis Y, Janssens D (2013) On the management and analysis of our LifeSteps. SIGKDD Explorations 15(1):23–32

Ruiz Vicente C, Freni D, Bettini C, Jensen CS (2011) Location-related privacy in geo-social networks. IEEE Internet Comput 15(3):20–27

Chapter 12
Hands-on with Hermes@Oracle MOD

Moving Object Database (MOD) engines enable us to process, manage and analyze mobility data. One of the already presented systems that satisfy these requirements is the Hermes MOD engine. Hermes provides MOD functionality to OpenGIS-compatible state-of-the-art Object-Relational DBMS. Currently, Hermes comes in two implementations; the first operates on top of Oracle DBMS (denoted as Hermes@Oracle) and the second operates on top of PostgreSQL (denoted as Hermes@Postgres). In Chap. 5, we studied the collection of abstract data types (ADT) and their corresponding operations of the Hermes@Oracle, which were defined, developed and provided as a data cartridge extending Oracle's SQL query language with MOD semantics. The present chapter contains a detailed description of a hands-on experience over a dataset of synthetic trajectories (produced by the Hermoupolis generator) moving in Attiki, Greece. There are several sample queries as well as analytical and mining procedures running over the Attiki dataset and showing Hermes@Oracle API in action. Regarding installation guidelines for Hermes, as well as the specific dataset used in the current hands-on, and the Hermoupolis generator for producing datasets with different properties, the interested reader is referred to Hermes@Oracle homepage.[1]

12.1 Introduction: The Hermes@Oracle Data Type System

As described in Chap. 5, Hermes MOD engine defines a series of spatiotemporal data types. In the current section we demonstrate the realization of these datatypes in the Oracle DBMS. Recall, that the lowest level of movement abstraction is a

---

[1] The version of Hermes demonstrated in this chapter is the one built upon Oracle DBMS. This implementation is available for downloading for research and educational purposes under Hermes license at URL: http://infolab.cs.unipi.gr/hermes.

N. Pelekis and Y. Theodoridis, *Mobility Data Management and Exploration*,
DOI 10.1007/978-1-4939-0392-4_12, © Springer Science+Business Media New York 2014

three-dimensional segment, meaning that a trajectory consists of a collection of such segments. Below, we provide an example for constructing a trajectory object that corresponds to a single segment moving on a straight line (recall the meaning of the 'PLNML_1' flag), from point (1,1) to point (2,2) between "2013-01-01 00:00:01" and "2013-01-01 00:00:59". Moreover, each such object carries its identifier and the identifier of the spatial reference system (SRID) with which it is represented (see the corresponding values '1, 2100' in line 8). In the usual case that a moving_point object includes many unit_moving_point objects, these are comma separated inside the moving_point_tab collection object.

Script S0. A sample trajectory data object.

```
1.   moving_point(
2.     moving_point_tab(
3.       unit_moving_point(
4.         tau_tll.d_period_sec(
5.           tau_tll.d_timepoint_sec(2013,1,1, 0,0,1),
6.           tau_tll.d_timepoint_sec(2013,1,1, 0,0,59) ),
7.         unit_function(1,1, 2,2, NULL, NULL, NULL, NULL, NULL, 'PLNML_1') )
8.     ), 1, 2100)
```

The produced dataset is hosted in 'attiki_mpoints' table in which trajectories are stored as moving_point objects, namely following a trajectory-oriented storage model (recall the discussion in Chaps. 3 and 4 about alternative models for trajectories). The list of attributes of 'attiki_mpoints' includes: <*object_id*, *traj_id*, *mpoint*>, where *object_id* corresponds to object's unique identifier, *traj_id* corresponds to a unique identifier of object's trajectory, and *mpoint* is the moving_point object.

Although 'attiki_mpoints' table contains the raw spatio-temporal data, the Hermoupolis generator (briefly presented in Chap. 3) also produces semantically-enriched information, which is synchronized with the raw counterpart. Following the approach presented in Chap. 9 (Sect. 9.3.1), here we present how this synchronization looks like inside the database. First of all, the raw information of 'attiki_mpoints' table is reformed to another table 'attiki_sub_mpoints' to reflect the partitioning of the trajectories to sub-trajectories. More specifically, for each trajectory of 'attiki_mpoints' table, 'attiki_sub_mpoints' table contains several tuples corresponding to the sub-trajectories of the episodes that the initial trajectory is partitioned. So, the 'attiki_sub_mpoints' list of attributes includes: <*o_id*, *traj_id*, *subtraj_id*, *sub_mpoint*>, where *sub_mpoint* is a moving_point object.

To exemplify the above, the following script presents an example of a sem_trajectory object that describes a 'middle-age male worker', represented with SRID 2100 (line 1), with <*o_id*, *traj_id*> equal to <1, 1> (line 13), which consists of two episodes: in the first episode that is a Stop, the object is sleeping at home, while in the second the object is driving on 'avenueE96'. The MBBs that approximate these portions of the trajectory are also kept together with the references (pointers) to the detailed raw data, which are tuples in the 'attiki_sub_mpoints' table. All the semantic trajectories of the Attiki dataset are hosted in the 'attiki_sem_trajs' table, which is a table of sem_trajectory objects.

Script S1. A sample semantic trajectory data object.

```
1.    sem_trajectory('middle-age male worker', 2100,
2.      sem_episode_tab(
3.        sem_episode('Stop', 'home', sleeping,
4.          sem_mbb(
5.            sem_st_point(1,2,tau_tll.D_Timepoint_sec(2013,1,1, 0,0,1)),
6.            sem_st_point(3,4, tau_tll.D_Timepoint_sec(2013,1,7, 0,0,1))),
7.          tlink),
8.        sem_episode('Move', 'avenueE96', 'driving',
9.          sem_mbb(
10.           sem_st_point(1,2,tau_tll.D_Timepoint_sec(2013,1,7, 0,0,2)),
11.           sem_st_point(3,4, tau_tll.D_Timepoint_sec(2013,1,8, 0,0,1))),
12.         tlink)),
13.       1,1)
```

12.2   The 'Attiki' Dataset

As mentioned earlier, the Attiki dataset (illustrated in Fig. 12.1) has been produced by Hermoupolis, which is a *pattern-aware synthetic trajectory generator* that produces annotated trajectories of moving objects in the form of sem_trajectory objects that follow given *Generalized Mobility Patterns* (GMP); in Fig. 12.1, each GMP is depicted with a different color. These GMP are represented in our case as semantic

Fig. 12.1 Overview of the Attiki dataset

trajectories, but with broad spatio-temporal components in order to simulate many specific semantic trajectories that share the same properties (e.g. all stop at certain regions where they perform the same activities, passing from them with the same order and similar transition times, while moving with similar means from one stop to another).

Apart from the set of semantic trajectories that correspond to profiles of movements, Hermoupolis takes as input a road network, which in our case is the road network of Attiki prefecture, and a set of *Points of Interest* (*PoI*). Both the road network and the PoI are retrieved from the OpenStreetMap open source repository. PoI are stored to the 'pois' table, which has the following attributes: an *id*, a *category* tag (e.g. pharmacy), a *name* and the point's spatial *location* as a geometry object. Given the above as input, Hermoupolis generates a set of network-constrained trajectories conforming to the requirements posed by GMP.

The output of Hermoupolis is stored in a relational table with the following scheme:

- id: The identifier of the object.
- GMPid: The identifier of the profile GMP that the object belongs to.
- episodeid: The identifier of the episode that corresponds to the reported position.
- Time: The time instant when the object reports its position.
- Simulation_time: The simulation time of the generator.
- X: The X ordinate of the reported position (in meters).
- Y: The Y ordinate of the reported position (in meters).
- Speed: The current speed of the object.
- defining_tag: The defining tag of the current episode (i.e. STOP or MOVE).
- episode_tag: Tag of the current episode (i.e. 'home').
- activity_tag: The activity tag that the object performs (i.e. 'sleeping').

Assuming that the output of Hermoupolis is stored to a table called 'attiki_reporter', then the following procedure creates and feeds the previously discussed tables that host the raw (i.e. 'attiki_sub_mpoints') as well as the semantic trajectories (i.e. 'attiki_sem_trajs'); the SRID is also necessary.

```
begin
   hermoupolis.raw2semtrajs('attiki_reporter', 'attiki_sem_trajs', 'attiki_sub_mpoints', 2100);
end;
```

Since in our case study we will use also the initial 'attiki_mpoints' table, we can reconstruct it by merging the corresponding sub-trajectories in the 'attiki_sub_mpoints' table, by executing the following script:

```
begin
   sem_reconstruct.submpointsmerging('attiki_sub_mpoints', 'attiki_mpoints');
end;
```

Once we have constructed a table containing a column of moving_point datatype we can also build a TB-tree index on the table (recall the discussion about TB-trees in Chaps. 4 and 5). For the case of the 'attiki_mpoints' table we execute the following script:

```
create index attiki_mpoints_tbtree on attiki_mpoints (mpoint) indextype is tbtree
    parameters('traj_id');
```

12.3 Extracting Dataset Statistics

Hermes users can easily extract many interesting statistics that facilitate the understanding of a dataset before the actual analysis takes place. The statistics are both global, in the sense that they describe the whole dataset, but they can also refer to specific objects (trajectories) in the MOD. To calculate either type of statistics for the raw Attiki dataset, we run the following script:

```
begin
    statistics.createdatasetstats('attiki_mpoints');
end;
```

The above procedure creates two tables, one for global statistics (called, 'attiki_mpo_global') and one for statistics per trajectory (called, 'attiki_mpo_pertraj'). Below we present some important statistics of the whole dataset, such as its spatio-temporal bounds, the number of points in the MOD, etc.:

	MINX	MAXX	MINY	MAXY	MINT	MAXT	SRID	NUMOFTRAJS
1	436010	504670	4173621	4240105	08-MAY-13 12.00.00.000000 AM ⋯	08-MAY-13 11.59.50.000000 PM ⋯	2100	4160

- Number of points of all trajectories: 1,431,162
- Minimum number of points in a trajectory: 10
- Maximum number of points in a trajectory: 1,805
- Average number of points per trajectory: 344
- Average duration of trajectories (seconds): 81
- Average length of trajectories (meters): 25,255

On the other hand, the information that we extract per trajectory, stored in 'attiki_mpo_pertraj' table, includes the starting and ending location of the trajectory as a geometry object, its MBB and measures like the duration, length, average speed, number of points and radius of gyration of the corresponding trajectory.

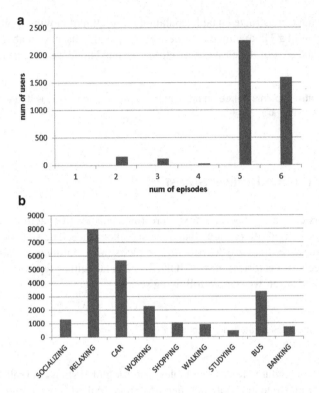

Fig. 12.2 Statistics of the Attiki dataset: (**a**) Number of users following specific number of episodes; (**b**) number of episodes per (either Stop or Move) activity

Let us now provide statistics about the semantic-aware representation of the trajectories. The total number of episodes is 23,736 (in detail, 9,937 Move and 13,799 Stop episodes), an average of 5.7 episodes per semantic trajectory. The profile set given as input to the Hermoupolis generator, includes six categories of Stop episode activities (i.e., purpose at destination site) and three of Move episodes (i.e., transportation means). In Fig. 12.2, we provide a diagrammatic representation of a few statistics extracted from the 'attiki_sem_trajs' table:

Furthermore, Table 12.1 presents some measures, like covered distance, duration and speed per episode type. Note that for Stops, the covered distance is zero, implying that the objects are absolutely still during the episode.

In the following two sections we provide a rich palette of example queries that demonstrate the query language of Hermes and reveal several potential analyses that could make use of such kind of functionality. The first section targets the raw spatio-temporal data, while the second the semantically-enriched MOD. We note that the description assumes that the reader is aware of the API of Hermes as this has been presented in Chaps. 5 and 9, respectively. Here, we will present auxiliary functionality that was not presented there.

Table 12.1 Statistics for episodes per episode type

	Stop	Move
Sum (average) of distance covered, in meters	0 (0)	103,879,945 (10,454)
Sum (average) of duration, in seconds	322,170,878 (23,347)	13,742,462 (1,383)
Average speed, in m/s	0	7.5

12.4 Querying the Raw GPS Part of the Dataset

12.4.1 Queries on Individual Trajectories

Let us first show how the 'attiki_mpoints' object-relational table looks like, where the raw mobility data reside. In this example a specific trajectory is retrieved as a moving_point object. As illustrated in the following figure, the whole trajectory is returned in a single tuple, in a single column of the relation.

```
SELECT m.object_id, m.traj_id, m.mpoint as trajectory
FROM attiki_mpoints m
WHERE m.traj_id = 24;
```

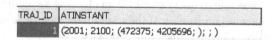

OBJECT_ID	TRAJ_ID	TRAJECTORY
24	24	(((((2013; 5; 8; 0; 0; 0); (2013; 5; 8; 0; 47; 5)); (473649; 4204623; 473649; 4204623; ; ; ; ; ; PLNML_1))); (((

Next, we find where a moving object was located at a specific timepoint. The query uses the *at_instant* method, which returns the SDO_GEOMETRY point object at which the moving object was at that particular time instant.

```
SELECT m.traj_id, m.mpoint.at_instant(tau_tll.d_timepoint_sec(2013,5,8,10,10,10))
       atinstant
FROM attiki_mpoints m
WHERE m.traj_id = 1;
```

TRAJ_ID	ATINSTANT
1	(2001; 2100; (472375; 4205696;); ;)

Next, we find the part of the trajectory that exists in a specific time period. By using the *at_period* method, the following query restricts the selected trajectory inside the given period. Obviously, the result is again of moving_point datatype.

```
SELECT m.mpoint.at_period(
  tau_tll.d_period_sec(
    tau_tll.d_timepoint_sec(2013,5,8,13,00,00),
    tau_tll.d_timepoint_sec(2013,5,8,15,00,00))) atperiod
  FROM attiki_mpoints m
WHERE m.traj_id = 12;
```

```
ATPERIOD
(((((2013; 5; 8; 13; 0; 0); (2013; 5; 8; 13; 25; 0)); (469798; 4204419; 469798; 4204419; ; ; ; ; ; PLNML_1)); (((2
```

Let us now present a query that cooperates with spatial-only functionality provided by Oracle Spatial. More specifically, the following (topological) query checks whether the position that a trajectory was located at a timepoint, interacts spatially with a polygon geometry. The specific input polygon is of rectangle type, thus, it is defined by its lower left (i.e., point (472500,4204720)) and upper right corner (i.e., point (472800,4204800)). The spatial topological test (resulting in TRUE in this example) takes place with the *relate* method of Oracle Spatial, using the 'ANYINTERACT' flag parameter.

```
SELECT SDO_GEOM.relate(
  SDO_GEOMETRY(2003, 2100, NULL, SDO_ELEM_INFO_ARRAY(1,1003,3),
              SDO_ORDINATE_ARRAY(472500,4204720, 472800,4204800)),
  'ANYINTERACT',
  m.mpoint.at_instant(tau_tll.d_timepoint_sec(2013,5,8,7,55,54)),0.001) relate
FROM attiki_mpoints m
WHERE m.traj_id = 24;
```

```
RELATE
TRUE
```

The following (directional) query checks if an object at a specific timepoint has a given point (of interest) on its east. If the test is true, then the query returns the route of the trajectory, which is a linestring geometry.

```
SELECT m.mpoint.route()
FROM attiki_mpoints m
WHERE m.traj_id = 24 AND
  m.mpoint.f_east(
    SDO_GEOMETRY(2001, 2100, SDO_POINT_TYPE(472900, 4204888,NULL),
    NULL, NULL),
    tau_tll.d_timepoint_sec(2013,5,8,7,54,51)) = 1;
```

In the following example, we compute the distance of a selected set of moving objects (i.e. {1, 12}) at a specific timepoint from a static point, which may be defined at query time (as in the current example) or retrieved from a PoI table.

```
SELECT m.traj_id, m.mpoint.f_distance(
  SDO_GEOMETRY(2001, 2100, SDO_POINT_TYPE(470000,4200000,NULL), NULL,
    NULL),  0.001,
  tau_tll.d_timepoint_sec(2013,5,8,7,59,52)
) distance_from_point
FROM attiki_mpoints m
WHERE m.traj_id IN (1, 12);
```

TRAJ_ID	DISTANCE_FROM_POINT
1	4569.25361793368
12	4000.25324198356

This query determines whether a trajectory existing in a predetermined set of object identifiers (e.g. {1, 1200}) is within a given distance (e.g. 2,000 m) from a certain location at a given timepoint.

```
SELECT m.object_id, m.mpoint.f_within_distance(2000,
  SDO_GEOMETRY(2001, 2100, SDO_POINT_TYPE(475948, 4203729,NULL),
    NULL, NULL),
  0.001,tau_tll.d_timepoint_sec(2013,5,8,9,10,00)) isit
FROM attiki_mpoints m
WHERE m.traj_id IN (1, 1200);
```

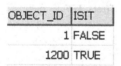

OBJECT_ID	ISIT
1	FALSE
1200	TRUE

The following query computes the angle (in degrees) between xx' axis and the directed segment that is created by a point residing on a trajectory at a specific time-point and another point of interest (i.e. (475948, 4203729)).

```
SELECT utilities.direction(
  m.mpoint.at_instant(tau_tll.d_timepoint_sec(2013,5,8,9,10,00)),
  SDO_GEOMETRY(2001, 2100, SDO_POINT_TYPE (475948, 4203729,NULL),
      NULL, NULL)) angle
FROM attiki_mpoints m
WHERE m.traj_id = 1200;
```

ANGLE
200.084728620755

The following example retrieves trajectories that start their movement from a given region. Note that this query can be resolved by a spatial range query on the ('attiki_mpo_pertraj') table that keeps statistics for each of the trajectories of our MOD, where the starting positions of them are maintained. This query ignores the existence of such a metadata table and resolves the query on-the-fly by retrieving the starting positions of the trajectories (note the use of the *f_initial* method that returns the starting location of a moving point). Although this is not the most efficient way to resolve this query, it is representative of the kind of cooperation that a MOD can have with a spatial DBMS.

```
DECLARE
  ret1 mp_array;
  sgeo SDO_GEOMETRY;
BEGIN
  sgeo := SDO_GEOMETRY(2003, 2100, NULL, SDO_ELEM_INFO_ARRAY(1,1003,3),
          SDO_ORDINATE_ARRAY(469358, 4205883, 471319, 4207570));
  visualizer.mbr2wkt(sgeo,'RANGE');
  SELECT m.mpoint BULK COLLECT INTO ret1
  FROM attiki_mpoints m
  WHERE SDO_GEOM.RELATE(m.mpoint.f_initial(), 'INSIDE', sgeo), 0.005) = 'INSIDE';
  visualizer.MovingPointTable2WKT(ret1,'TRAJS',null);
END;
```

In detail, the core of the above script is the SQL query. The script further exhibits a simple way that Hermes provides to visualize query results. More specifically, a package called Visualizer allows a user to transform Oracle Spatial and Moving Point objects to either KML or Well-Known-Text (WKT) files, which can be directly visualized by various GIS tools (e.g. Google Earth and OpenJump, respectively). The user only needs to invoke the appropriate method

Fig. 12.3 Trajectories
starting their movement
inside a rectangular region

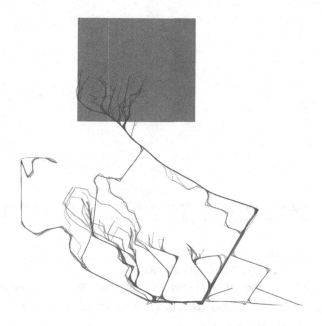

that visualizes a specific datatype (i.e. 'mbr2WKT' for rectangles and 'MovingPointTable2WKT' for arrays of moving_point objects) and give as input the object to be visualized and a string that will name the resulting KML or WKT file. In this example, we first create a file 'RANGE.wkt' that contains the rectangular area. Then, we create the 'TRAJS.wkt' file, which contains the result set of the trajectories. Figure 12.3 visualizes both as two separate layers in OpenJump open-source GIS tool.

The following example is a simple PL/SQL script that starts by declaring a series of parameters of various type and then it exports a rectangle region and a trajectory as WKT files for visualization. Then, the script uses the *f_intersection* method to find the part of the trajectory that resides within the previous region, while it subsequently finds the exact point, the time instance (see *f_enter* method) and the speed (see *f_speed* method) that the trajectory enters the region. All the steps of the script visualize the involved objects. (Please take note that the particular script does not include all visualization statements that result in the following figures.) Most of them result in WKT files (those that their last three characters of their method name is 'wkt'). By changing the last three characters of the function name of the visualizer (i.e. from 'wkt' to 'kml' and vice versa), we may exchange the visualization mode. The visualization outcome of this query in 'kml' regarding the entering point with a Google Earth placemark (not available in WKT) is omitted (Fig. 12.4).

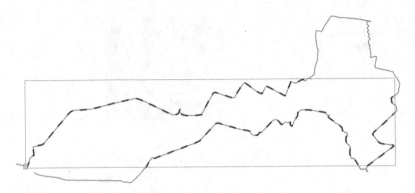

Fig. 12.4 The route of a trajectory (*solid line*), its restriction (*dashed line*) inside a rectangular region, and a placemark of the first entering point

```
DECLARE
    geom MDSYS.SDO_GEOMETRY:=
            SDO_GEOMETRY(2003, 2100, NULL, SDO_ELEM_INFO_ARRAY(1,1003,3),
                SDO_ORDINATE_ARRAY(465700,4201300, 477900,4204100));
    tp tau_tll.d_timepoint_sec;
    enterpoint MDSYS.SDO_GEOMETRY;
    speed NUMBER;
    id PLS_INTEGER:=524;
    mpoints mp_array;
    mpoints2 mp_array;
BEGIN
    VISUALIZER.geom_to_wkt(geom, 'RECTANGLE');
    SELECT m.mpoint bulk collect into mpoints
    FROM attiki_mpoints m
    WHERE m.traj_id = id;
    VISUALIZER.movingpointtable2wkt(mpoints, 'MOVING_POINT_WHOLE.wkt',null);
    SELECT m.mpoint.f_intersection(geom, 0.001) bulk collect into mpoints2
    FROM attiki_mpoints m
    WHERE m.traj_id = id;
    IF mpoints2.count>0 THEN
        VISUALIZER.movingpointtable2wkt(mpoints2, 'MOVING_POINT_PART.wkt',null);
```

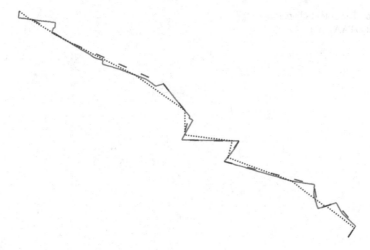

Fig. 12.5 A trajectory (*solid line*) and its simplified counterpart produced by the TD_TR (*dashed line*) and the BOPW_TR (*dotted line*) algorithms

```
tp := mpoints(mpoints.last).f_enter(geom);
enterpoint := mpoints(mpoints.last).at_instant(tp);
speed := mpoints(mpoints.last).f_speed(tp);

VISUALIZER.Placemark2KML(enterpoint, 8307, 'ENTER.kml',
 'WHEN, WHERE and with WHAT SPEED DID TRAJECTORY ' || id ||
    ' ENTER the AREA?', 'HERE @ ' || tp.to_string() ||
    ' WITH SPEED=' || TO_CHAR(speed) || ' m/sec');
ELSE
   dbms_output.put_line('THERE IS NO INTERSECTION!');
   END IF;
END;
```

In the following example, a trajectory is simplified using the TD-TR algorithm, which is an extension of Douglas-Peucker algorithm that takes into account the temporal dimension as it uses the Synchronous Euclidean Distance (SED), discussed in Chap. 3. The same trajectory is subsequently compressed in an online fashion using BOPW-TR algorithm, a variant of TD-TR algorithm. The distance threshold for parameterizing the algorithms is set to 100 m. All three versions of the trajectory are visualized in Fig. 12.5, where one can note that TD-TR compresses more effectively than BOPW-TR, due to the online computation approach of the latter, which however negatively affects the efficiency of the simplification.

Fig. 12.6 The potential area
of activity (PAA) of a
trajectory

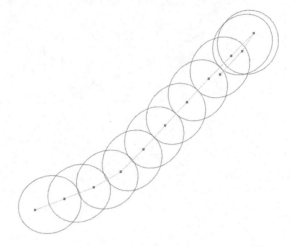

```
declare
    mp moving_point;
    tmp moving_point;
    mpoints mp_array; bmp moving_point;
begin
    select m.mpoint into mp from attiki_mpoints m where m.traj_id = 45;
    mpoints:=mp_array(mp);
    visualizer.movingpointtable2wkt(mpoints,'moving_point.wkt',null);

    select td_tr(mp, 100) into tmp from dual;
    mpoints:=mp_array(tmp);
    visualizer.movingpointtable2wkt(mpoints,'simplified_moving_point.wkt',null);

    select bopw_tr(mp, 100) into bmp from dual;
    mpoints:=mp_array(bmp);
    visualizer.movingpointtable2wkt(mpoints,'online_simplified_moving_point.wkt',null);
end;
```

In the following example, we compute the potential activity area (PAA) of a
trajectory and we project it on the map (recall that PAA was discussed in Chap. 4).
In this particular case, the PAA is calculated for every segment of a trajectory based
on the maximum speed of the object during its whole lifespan. As such, for each
segment of movement we calculate the corresponding ellipsis, as illustrated in
Fig. 12.6. Having computed the PAA, it is trivial to join with a table of PoI and find
the places wherefrom an object has passed.

Fig. 12.7 Trajectories with ids 3590 (*light grey dashed line*) and 3657 (*dark grey dashed line*) came close to each other (closer than 10 m) 22 times. Note that after resampling both trajectories consist of 163 three-dimensional segments

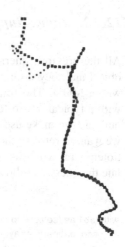

```
declare
    paa sdo_geometry;
begin
    select m.mpoint.potential_activity_area() into paa
    from attiki_mpoints m
    where m.traj_id=2935;
    visualizer.geom_to_wkt(sdo_cs.transform(paa, 2100) , 'paa_of_traj_2935');
end;
```

The subsequent example demonstrates a simple way to measure the similarity between two trajectories by counting how many times two moving objects came close to each other in terms of an Euclidean distance threshold (i.e. 10 m). Recall that the proximity test applied by the *number_of_times_close* method (discussed in Chap. 5) between the two trajectories is evaluated only at a predefined sequence of timestamps (Fig. 12.7).

```
with tab as (select m.mpoint from attiki_mpoints m where m.traj_id=3657)
select m.mpoint.number_of_times_close(t.mpoint, 10, 0.5)
from attiki_mpoints m,tab t
where m.traj_id=3590;
```

M.MPOINT.NUMBER_OF_TIMES_CLOSE(T.MPOINT,10,0.5)
22

12.4.2   Index-Supported Queries

All the previous queries are *atomic*, in the sense that they operate per moving object and they do not make use of an index, as the query is not applied to the whole MOD. The following examples show index-supported queries, starting with a typical spatio-temporal range query. Note that in absence of an index (in our case we make use of the TB-tree) in order to resolve a range query (actually we want to retrieve the parts of the trajectories that are contained in a given spatiotemporal window) with functionality previously presented, we should formulate the query as follows:

```
with tab1 as (select sdo_geometry(2003, 2100, null, sdo_elem_info_array(1,1003,3),
     sdo_ordinate_array(475520,4202902, 478320,4204402)) geom from      dual),
  tab2 as (select tau_tll.d_period_sec(tau_tll.d_timepoint_sec(2013,5,8,8,15,00),
     tau_tll.d_timepoint_sec(2013,5,8,16,50,50)) tp from dual)
select m.mpoint.at_period(tab2.tp).f_intersection(tab1.geom, 0.01)
from attiki_mpoints m, tab1, tab2
where m.mpoint.at_period(tab2.tp).f_intersection(tab1.geom, 0.01) is not null;
```

The following syntax takes advantage of the index, as such improving drastically the processing time. The query result, illustrated in Fig. 12.8, consists of 447 trajectories falling inside the given spatio-temporal box.

The subsequent query returns the geometries of entering and leaving positions of

```
select * from table(
   tbfunctions.range(sdo_geometry(2003, 2100, null, sdo_elem_info_array(1,1003,3),
                     sdo_ordinate_array(475520,4202902, 478320,4204402)),
             tau_tll.d_period_sec(tau_tll.d_timepoint_sec(2013,5,8,8,15,00),
                          tau_tll.d_timepoint_sec(2013,5,8,16,50,50)),
   2100, 'attiki_mpoints_tbtree_non_leaf ', 'attiki_mpoints_tbtree_leaf '));
```

the trajectories that crossed a given geometry within a specific time window. The trajectories on which the *get_enter_leave_points* is applied, are filtered first by the *topological* index-supported method, which in this particular case uses the mask 'enter_leave'. As the select clause of the query includes the output of the *get_enter_leave_points*, the final result (depicted in Fig. 12.9) consists of a multi-point collection geometry per trajectory, as each moving object may cross a geometry many times.

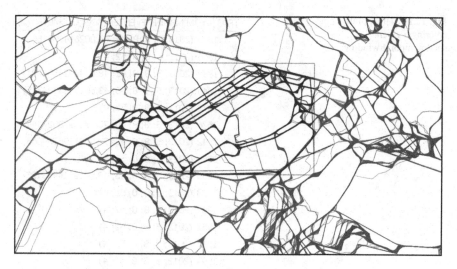

Fig. 12.8 Trajectories inside a spatio-temporal range

```
M.MPOINT.GET_ENTER_(206405.407)))
(2005; 2100; ; (1; 1; 2; ); (484910.316; 4205560.98677966; 484489.506; 4205483.14458333; ; ; ; ; ; ; ; ; ; ; ; ; ; ; ; ; ; ; ; ; ; ; ; ; ; ; ; ;
(2005; 2100; ; (1; 1; 2; ); (484543; 4205920; 484489.506; 4205981.04235514; ; ; ; ; ; ; ; ; ; ; ; ; ; ; ; ; ; ; ; ; ; ; ; ; ; ; ; ; ; ; ;
(2005; 2100; ; (1; 1; 2; ); (484910.316; 4205560.98677966; 484489.506; 4205483.14458333; ; ; ; ; ; ; ; ; ; ; ; ; ; ; ; ; ; ; ; ; ; ; ; ; ; .
(2005; 2100; ; (1; 1; 2; ); (484729.285726027; 4205405.823; 484720.833931372; 4205405.823; ; ; ; ; ; ; ; ; ; ; ; ; ; ; ; ; ; ; ; ; ; ; ; ; ;
(2005; 2100; ; (1; 1; 2; ); (484785; 4205962; 484489.506; 4205983.4368764; ; ; ; ; ; ; ; ; ; ; ; ; ; ; ; ; ; ; ; ; ; ; ; ; ; ; ; ; ; ; ;
(2005; 2100; ; (1; 1; 2; ); (484910.316; 4205520.10709677; 484728.485413333; 4205405.823; ; ; ; ; ; ; ; ; ; ; ; ; ; ; ; ; ; ; ; ; ; ; ; ; ; ; ;
(2005; 2100; ; (1; 1; 2; ); (484910.316; 4205561.518; 484489.506; 4205482.62718056; ; ; ; ; ; ; ; ; ; ; ; ; ; ; ; ; ; ; ; ; ; ; ; ; ; ; ; ; ;
(2005; 2100; ; (1; 1; 2; ); (484628; 4206193; 484489.506; 4206059.2056; ; ; ; ; ; ; ; ; ; ; ; ; ; ; ; ; ; ; ; ; ; ; ; ; ; ; ; ; ; ; ; ; ;
(2005; 2100; ; (1; 1; 2; ); (484729.285726027; 4205405.823; 484720.833931372; 4205405.823; ; ; ; ; ; ; ; ; ; ; ; ; ; ; ; ; ; ; ; ; ; ; ; ; ; ; ;
(2005; 2100; ; (1; 1; 2; ); (484817; 4206129; 484489.506; 4206071.91495868; ; ; ; ; ; ; ; ; ; ; ; ; ; ; ; ; ; ; ; ; ; ; ; ; ; ; ; ; ; ; ; ;
(2005; 2100; ; (1; 1; 2; ); (484699; 4205438; 484489.506; 4205483.31468056; ; ; ; ; ; ; ; ; ; ; ; ; ; ; ; ; ; ; ; ; ; ; ; ; ; ; ; ; ; ; ; ; ;
(2005; 2100; ; (1; 1; 4; ); (484704; 4205478; 484910.316; 4205561.68749153; 484910.316; 4205601.59067717; 484719.002310679; 4205405.823;
(2005; 2100; ; (1; 1; 2; ); (484597; 4205836; 484489.506; 4205995.89888525; ; ; ; ; ; ; ; ; ; ; ; ; ; ; ; ; ; ; ; ; ; ; ; ; ; ; ; ; ; ; ; ;
(2005; 2100; ; (1; 1; 2; ); (484597; 4205836; 484489.506; 4205995.89888525; ; ; ; ; ; ; ; ; ; ; ; ; ; ; ; ; ; ; ; ; ; ; ; ; ; ; ; ; ; ; ; ;
(2005; 2100; ; (1; 1; 2; ); (484597; 4205836; 484489.506; 4205995.89888525; ; ; ; ; ; ; ; ; ; ; ; ; ; ; ; ; ; ; ; ; ; ; ; ; ; ; ; ; ; ; ; ;
(2005; 2100; ; (1; 1; 2; ); (484597; 4205836; 484489.506; 4205995.89888525; ; ; ; ; ; ; ; ; ; ; ; ; ; ; ; ; ; ; ; ; ; ; ; ; ; ; ; ; ; ; ; ;
(2005; 2100; ; (1; 1; 2; ); (484729.285726027; 4205405.823; 484720.833931372; 4205405.823; ; ; ; ; ; ; ; ; ; ; ; ; ; ; ; ; ; ; ; ; ; ; ; ; ; ; ;
(2005; 2100; ; (1; 1; 2; ); (484543; 4205920; 484489.506; 4205981.04235514; ; ; ; ; ; ; ; ; ; ; ; ; ; ; ; ; ; ; ; ; ; ; ; ; ; ; ; ; ; ; ; ;
(2005; 2100; ; (1; 1; 2; ); (484746; 4205430; 484720.620074074; 4205405.823; . . . . . . . . . . . . . .
```

Fig. 12.9 Points entering or leaving a given spatiotemporal window

The following query is a variant of the previous, as it returns the timepoints (see Fig. 12.10) that trajectories enter the given spatio-temporal box. This time note the use of the 'enter' mask in the *topological* method.

Fig. 12.10 Timepoints
entering a given
spatiotemporal window

TRAJ_ID	M.MPOINT.F_ENTER(SDO_GEOMETRY(2(
1800	(2013; 5; 8; 8; 41; 44.0020141601563)
1806	(2013; 5; 8; 0; 52; 4)
1833	(2013; 5; 8; 8; 41; 48.8349914550781)
1840	(2013; 5; 8; 8; 40; 43.6829833984375)
1872	(2013; 5; 8; 0; 52; 4)
2526	(2013; 5; 8; 8; 40; 20.3789978027344)
2542	(2013; 5; 8; 8; 42; 7.25601196289063)
2562	(2013; 5; 8; 0; 52; 4)
2565	(2013; 5; 8; 8; 45; 18.5059814453125)
2573	(2013; 5; 8; 0; 52; 4)
1902	(2013; 5; 8; 0; 52; 4)
1957	(2013; 5; 8; 0; 52; 4)
1971	(2013; 5; 8; 0; 52; 4)
1994	(2013; 5; 8; 0; 52; 4)
2009	(2013; 5; 8; 0; 52; 4)
2021	(2013; 5; 8; 0; 52; 4)
2069	(2013; 5; 8; 8; 40; 43.6829833984375)
2080	(2013; 5; 8; 0; 52; 4)
2082	(2013; 5; 8; 0; 52; 4)
2086	(2013; 5; 8; 0; 52; 4)
2098	(2013; 5; 8; 0; 52; 4)
2105	(2013; 5; 8; 8; 41; 58.4250183105469)

```
select m.traj_id, m.mpoint.get_enter_leave_points(
      sdo_geometry(2003, 2100, null, sdo_elem_info_array(1,1003,3),
            sdo_ordinate_array(484489.506,4205405.823, 484910.316,4206405.407)))
from attiki_mpoints m
where m.traj_id in (
    select distinct * from table (
      tbfunctions.topological(
          sdo_geometry(2003, 2100, null, sdo_elem_info_array(1,1003,3),
            sdo_ordinate_array(484489.506,4205405.823, 484910.316,4206405.407)),
          tau_tll.d_period_sec(tau_tll.d_timepoint_sec(2013,5,8,7,30,00),
                        tau_tll.d_timepoint_sec(2013,5,8,16,00,00)),
          'enter_leave',
          2100, 'attiki_mpoints_tbtree_non_leaf ', 'attiki_mpoints_tbtree_leaf ')));
```

```
select m.traj_id, m.mpoint.f_enter(
    sdo_geometry(2003, 2100, null, sdo_elem_info_array(1,1003,3),
        sdo_ordinate_array(484489.506,4205405.823, 484910.316,4206405.407)))
from attiki_mpoints m
where m.traj_id in (
    select distinct * from table (
        tbfunctions.topological(
            sdo_geometry(2003, 2100, null, sdo_elem_info_array(1,1003,3),
                sdo_ordinate_array(484489.506,4205405.823, 484910.316,4206405.407)),
            tau_tll.d_period_sec(tau_tll.d_timepoint_sec(2013,5,8,7,30,00),
                                 tau_tll.d_timepoint_sec(2013,5,8,16,00,00)),
        'enter',
        2100, 'attiki_mpoints_tbtree_non_leaf ', 'attiki_mpoints_tbtree_leaf ')));
```

The following two queries demonstrate the available NN queries in Hermes, as these have been presented in Chap. 5. More specifically, the following SQL applies an Incremental Point k-NN query, which results in retrieving the $k = 3$ nearest three-dimensional segments (shown as *tbpoints* datatypes in Fig. 12.11, along with the corresponding trajectory identifiers) to a given spatiotemporal box, defined as an entry of a TB-tree node (i.e. tbmovingobjectentry). The third integer that is part of the construction statement of a *tbpoint* is simply the UNIX time of the corresponding timestamp.

```
select * from table(
    tbfunctions.incrpointnn(
    tbmovingobjectentry(null, tbpoint(tbx(468000, 4205500, 212234821200)),
                              tbpoint(tbx(471000, 4207090, 212234832000))),
    3,
    'attiki_mpoints_tbtree_leaf ', 'attiki_mpoints_tbtree_non_leaf '));
```

Similarly the subsequent query is an example of an Incremental Trajectory k-NN query, which, in contrast to the previous one, retrieves the nearest segments with respect to a given trajectory (Fig. 12.12).

```
select * from table(
    tbfunctions.incrtrajectorynn(
    (select m.mpoint from attiki_mpoints m where traj_id = 2),
    3,
    'attiki_mpoints_tbtree_leaf', 'attiki_mpoints_tbtree_non_leaf'));
```

ID	P1	P2
372	((468412; 4205260; 212234825740))	((468412; 4205272; 212234825745))
355	((468412; 4205257; 212234825770))	((468412; 4205270; 212234825774))
335	((468412; 4205257; 212234825767))	((468412; 4205269; 212234825770))

Fig. 12.11 The 3-NN segments (as three-dimensional boxes) with respect to a given three-dimensional box

ID	P1	P2
3836	((470080; 4204919; 212234805615))	((469954; 4204959; 212234805619))
3810	((469285; 4204802; 212234858391))	((469285; 4204802; 212234860790))
3781	((469285; 4204802; 212234858391))	((469285; 4204802; 212234860790))

Fig. 12.12 The 3-NN segments (as three-dimensional boxes) to a given trajectory

12.5   Querying the Semantically-Enriched Part of the Dataset

One could also pose queries targeting at the semantically-enriched trajectory database (STD). The first of the example demonstrates a way to visualize a semantic trajectory. In detail, the *semtrajectory2kml* function visualizes either the approximated sub-trajectories of the episodes (namely, their MBR) and/or the raw sub-trajectories, and/or the corresponding tags depending on a triplet of flag parameters. In the specific example, all flags are set, so Fig. 12.13 provides a complete view of the stored data for the selected semantic trajectory. Note that the textual information is depicted attached to a placemark that is located to the center of the episode's MBB.

```
declare
    semtraj sem_trajectory;
begin
    select value (s) into semtraj
    from attiki_sem_trajs s
    where o_id = 487 and semtraj_id = 487;
    visualizer.semtrajectory2kml (semtraj, 'true', 'true', 'true');
end;
```

The following query counts the number of Stops and Moves for all semantic trajectories (Fig. 12.14).

```
select o_id, semtraj_id,
    value (t).num_of_stops () num_of_stops,
    value (t).num_of_moves () num_of_moves,
    (value (t).num_of_stops () + value (t).num_of_moves () ) as num_of_episodes
from attiki_sem_trajs t
order by 4 desc;
```

Fig. 12.13 Visualizing a semantic trajectory

O_ID	SEMTRAJ_ID	NUM_OF_STOPS	NUM_OF_MOVES	NUM_OF_EPISODES
2820	2820	4	3	7
2821	2821	4	3	7
2822	2822	3	3	6
2823	2823	4	3	7
2825	2825	4	3	7
2826	2826	4	3	7
2827	2827	4	3	7
2828	2828	4	3	7
2830	2830	4	3	7
2832	2832	4	3	7
2833	2833	3	3	6
2834	2834	4	3	7
2835	2835	4	3	7
2836	2836	3	3	6
2837	2837	4	3	7

Fig. 12.14 Number of Stops and Moves per semantic trajectory

DURATION	DEFINING_TAG	ACTIVITY_TAG	NUM_OF_EPISODES
39608	STOP	RELAXING	2
9900	STOP	SOCIALIZING	1
28800	STOP	BANKING	1

Fig. 12.15 Duration of Stop episodes per activity

DURATION	DEFINING_TAG	ACTIVITY_TAG
1193	MOVE	CAR
1161	MOVE	CAR

Fig. 12.16 Move episodes travelling less time than average duration of Stop episodes whose activity is working

The following query summates the durations of Stop episodes per activity, for a given semantic trajectory (Fig. 12.15).

```
select  sum (value(s).duration().m_value ) duration,
        defining_tag, activity_tag, count (defining_tag) num_of_episodes
from table (select t.episodes_with ('stop')
        from attiki_sem_trajs t
        where t.o_id = 1045 and t.semtraj_id = 1045) s
group by defining_tag, activity_tag;
```

The following query demonstrates how one can filter episodes with multiple tags. More specifically, the query identifies those Move episodes whose duration is less than the average duration of Stop episodes, whose activity is *'working'* (Fig. 12.16).

```
select value(ext_s).duration().m_value duration, defining_tag, activity_tag
from table
   (select t.episodes_with ('move')
   from attiki_sem_trajs t where t.o_id    = 3125 AND t.semtraj_id = 3125) ext_s
   where (value(ext_s).duration().m_value ) <
      (select avg (value(s).duration().m_value)
      from table
         (select t.episodes_with ('working')
         from attiki_sem_trajs t
         where t.o_id = 3125 AND t.semtraj_id = 3125 AND
         t.episodes_with ('stop') is not null) s
      group by defining_tag);
```

DEFINING_TAG	EPISODE_TAG	ACTIVITY_TAG	MBB	TLINK
STOP	BANK	WORKING	((471652; 4200247; (2013; 5; 8; 9; 0; 0)); (471652; 4200247; (2013; 5; 8; 17; 0; 0)))	(REFERENCE)

Fig. 12.17 Stop episodes with 'working' tag inside a spatiotemporal window

DEFINING_TAG	EPISODE_TAG	ACTIVITY_TAG	MBB	TLINK
STOP	HOME	RELAXING	((470331; 4206228; (2013; 5; 8; 0; 0; 0)); (470331; 4206228; (2013; 5; 8; 7; 48; 36)))	(REFERENCE)
STOP	BANK	WORKING	((471652; 4200247; (2013; 5; 8; 9; 0; 0)); (471652; 4200247; (2013; 5; 8; 17; 0; 0)))	(REFERENCE)
STOP	HOME	RELAXING	((469285; 4204802; (2013; 5; 8; 17; 20; 0)); (469285; 4204802; (2013; 5; 8; 23; 59; 50)))	(REFERENCE)

Fig. 12.18 Episodes with either a 'stop' or a 'working' tag inside a spatiotemporal window

Let us now confine a semantic trajectory in the temporal dimension as well as by filtering those episodes w.r.t. their textual component. In detail, we restrict a given semantic trajectory inside a spatiotemporal box and then we retrieve only the Stop episodes that the user was working (Fig. 12.17).

```
select * from table (
   select b.confined_in (
                 sdo_geometry(2003,2100,null,sdo_elem_info_array(1,1003,3),
                              sdo_ordinate_array(469115,4200000, 472377,4206200)),
                 tau_tll.d_period_sec (tau_tll.d_timepoint_sec (2013, 5, 8, 08, 00, 00 ),
                                       tau_tll.d_timepoint_sec (2013, 5, 8, 23, 59, 00 )),
                 'stop' ).episodes_with ('working')
from attiki_sem_trajs b
where b.o_id = 3125
and b.semtraj_id = 3125);
```

The following query is a variant of the previous one that restricts a given semantic trajectory inside a temporal period and then returns episodes whose tags are either Stop or Working (Fig. 12.18).

```
select * from table (
   select b.confined_in (
                 null,
                 tau_tll.d_period_sec (tau_tll.d_timepoint_sec (2013, 5, 8, 08, 00, 00 ),
                                       tau_tll.d_timepoint_sec (2013, 5, 8, 23, 59, 00 )),
                 'stop+working' ).episodes
from attiki_sem_trajs b
where b.o_id = 3125
and b.semtraj_id = 3125 );
```

The above-described examples exhibit some of the object methods of the datatype for semantic trajectories. Below we provide further examples that aim to retrieve a subset of the table storing semantic trajectories. As such, to speed up queries we build a specialized index (namely the STB-tree presented in Chap. 9) on the 'attiki_sem_trajs' table. To build a STB-tree we need to execute the following procedure. Obviously, the first parameter is the name of the created index, which will be given as parameter to many operators, so as to let the operator know which of the (potential many) indices will be used, while the second is the name of the table containing the semantic trajectories, upon which the index will be built.

```
begin
  std.create_sem_tbtree('attiki_stbtree', 'attiki_sem_trajs');
end;
```

Given such an index, the following example applies a temporal range query to count for each valid semantic trajectory inside the specified temporal period, how many Move episodes (and with what total duration) each object has (Fig. 12.19).

```
select deref (tlink).o_id o_id,  count (tlink) total_moves,
     sum (value(s).duration().m_value) total_duration
from table
     (select std.stb_range_episodes ('move',
          tau_tll.d_period_sec (tau_tll.d_timepoint_sec (2013, 5, 8, 8, 01, 00 ),
                              tau_tll.d_timepoint_sec (2013, 5, 8, 16, 59, 00 ) ), 'attiki_stbtree' )
     from dual) s
group by deref (tlink).o_id
order by 1;
```

The following query finds the average duration of Stop episodes overlapping with a spatio-temporal window. The long duration, shown in Fig. 12.20, rather implies that the corresponding Stops are either for work at a business district or for relaxing at a regional area.

```
select sum(value(s).duration().m_value) / count(deref(tlink).traj_id)
   avg_mbb_stop_duration
from table
   (select std.stb_range_episodes ('stop',
       mdsys.sdo_geometry (2003, 2100, null, mdsys.sdo_elem_info_array (1, 1003, 3 ),
       mdsys.sdo_ordinate_array (470000, 4200000, 474000, 4205000 ) ),
       tau_tll.d_period_sec (tau_tll.d_timepoint_sec (2013, 5, 8, 7, 50, 00),
                            tau_tll.d_timepoint_sec(2013, 5, 8, 9, 10, 00) ), 'attiki_stbtree' )
   from dual) s;
```

Fig. 12.19 Number of Move
episodes and their durations
inside a given period

O_ID	TOTAL_MOVES	TOTAL_DURATION
4242	1	1193
4243	1	1193
4244	1	1193
4245	1	1193
4246	1	1173
4247	1	1193
4248	1	1193
4249	1	1193
4250	2	3541
4252	2	3535
4253	2	3545
4255	1	1702
4256	2	3544
4257	2	3549
4258	2	3532

Fig. 12.20 Average duration
of Stop episodes in a
spatio-temporal box

AVG_MBB_STOP_DURATION
25049.696969697

The previous examples use only the approximate spatio-temporal and the semantic information as they do not require the detailed spatio-temporal information available (via the T-link pointer) at the raw MOD. The following example is a cross-over spatio-temporal range query with an additional textual constraint that is resolved using the STB-tree (filter step) and a subsequent temporal restriction of the sub-trajectories of the resulting Move episodes (refinement step).

```
select deref(tlink).sub_mpoint sub_mpoint,
        deref(tlink).sub_mpoint.at_period(tau_tll.d_period_sec(
                tau_tll.d_timepoint_sec(2013, 5, 8, 7, 50, 00),
                    tau_tll.d_timepoint_sec(2013, 5, 8, 14, 10, 00)))
                        restricted_sub_mpoint,
        deref (tlink).o_id o_id
from table
   (select std.stb_range_episodes('move',
        mdsys.sdo_geometry (2003, 2100, null, mdsys.sdo_elem_info_array (1, 1003, 3),
        mdsys.sdo_ordinate_array(470000, 4200000, 474000, 4205000) ),
        tau_tll.d_period_sec (tau_tll.d_timepoint_sec (2013, 5, 8, 7, 50, 00),
                tau_tll.d_timepoint_sec(2013, 5, 8, 14, 10, 00)), 'attiki_stbtree')
        from dual) s
where
        deref(tlink).sub_mpoint.at_period(tau_tll.d_period_sec(
                tau_tll.d_timepoint_sec(2013, 5, 8, 7, 50, 00),
                tau_tll.d_timepoint_sec(2013, 5, 8, 14, 10, 00))) is not null;
```

Fig. 12.21 Distribution of
objects with respect to
transportation mode that start
from locations annotated as
'home'

ACTIVITY_TAG	COUNT(MOV_OBJ)
CAR	1218
WALKING	460
BUS	697

Fig. 12.22 Trajectory ids
following the pattern
'bus*bank*cafe'

COLUMN_VALUE
1207
701
718
979
1221
1118
833
1011
982
787
1009

The following example demonstrates the variant of the spatial-temporal-textual range query that identifies patterns of the form "from-to-via". More specifically, the example computes the distribution of objects per transportation mode that start from '*Home*' inside a given temporal period. The result is depicted in Fig. 12.21.

```
select activity_tag, count(mov_obj)
from (select deref(tlink).o_id mov_obj, activity_tag
     from table(std.stb_from_to_via(
         sem_episode('stop',   'home',null,null,null),
         sem_episode('stop',    null,null,null,null),
         sem_episode('move',null,null,
          sem_mbb(
          sem_st_point(470000, 4200000, tau_tll.d_timepoint_sec(2013, 5, 8, 7, 50, 00)),
          sem_st_point(474000, 4205000,tau_tll.d_timepoint_sec(2013, 5, 8, 9, 50, 00)))
           ,null),
          'attiki_stbtree')) t)
group by activity_tag;
```

Finally, we demonstrate an interesting query that searches for semantic trajectories that follow a specific pattern, which is represented as a sequence of tags, possibly separated by wildcard characters (e.g. '*' implying any sub-sequence of tags). For instance, the following query retrieves semantic trajectories (their identifiers are depicted in Fig. 12.22) following the pattern 'bus*bank*cafe', namely objects travelling by bus, then stopping to a bank and subsequently to a café, while the pattern matches objects that may have one or more intermediate stops (i.e. activities) and/or transportation modes between them.

```
select std.stb_patterns('bus*bank*cafe', 'attiki_stbtree') from dual;
```

12.6 Trajectory Warehousing and OLAP in Hermes@Oracle

Hermes MOD engine provides the functionality to create a trajectory data warehouse (TDW), to load it with data extracted and transformed from a MOD and to execute a number of OLAP operations. Let us demonstrate the building process of a TDW that follows the subsequent scheme (a simplified version of the one presented in Chap. 6) (Fig. 12.23).

In order to build a TDW for the 'attiki_mpoints' we use the createtdw('attiki') procedure available in the TDW package. The 'attiki' parameter is actually a prefix string that it is used to create the TDW scheme with appropriate naming. In detail, this call will produce the dimension tables "ATTIKI_RECTANGLE", "ATTIKI_TIME_PERIODS" and the fact table " ATTIKI_FACTTBL".

```
begin
    tdw.createtdw('attiki');
end;
```

To load the dimension tables, two procedures are used, namely the SPLITSPACE and SPLITTIME. These are necessary to split the spatial and temporal dimensions into predefined spatial areas and temporal periods. The SPLITSPACE procedure

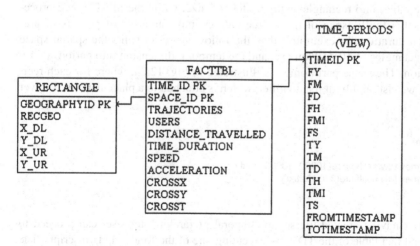

Fig. 12.23 A star scheme for a trajectory data warehouse

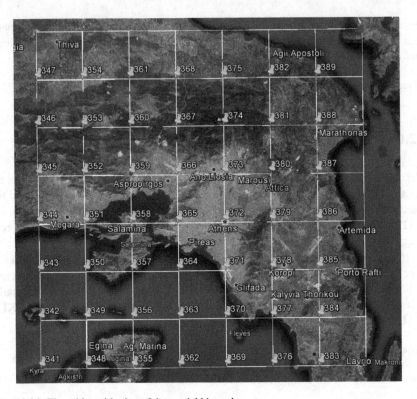

Fig. 12.24 The grid partitioning of the spatial hierarchy

splits the spatial region of the MBR that approximates all the trajectories of the
source table, into rectangles with predefined size, while the SPLITTIME procedure splits the lifespan of the source dataset into a number of periods of predefined duration. To exemplify this, the following script splits the spatial space
into rectangles of 10×10 km size and the temporal dimension into periods of 1 h
duration. The space partitioning is illustrated in Fig. 12.24, where for each rectangle we also highlight its identifier, which is attached to a placemark at its lower
left point.

```
declare
begin
    hermes.tdw.splitspace(10000, 10000, 'attiki');
    hermes.tdw.splittime(3600,'attiki');
end;
```

After having created the spatio-temporal hierarchies, the user can proceed to
load the fact table of the TDW by executing one of the following two scripts. The
first script corresponds to cell-oriented ETL approach, while the second to the

```
AGGREGATIONS.AGGREGATED_NUM_OF_TRAJS(370,378,4645,4650,'ATTIKI')
                                                              1748
```

Fig. 12.25 Roll-up over the spatial and temporal dimension for the distinct number of trajectories measure

trajectory-oriented ETL approach, both described in Chap. 6. The *calculateauxiliary_cl* procedure computes the last three measures of the fact table (i.e. the number of trajectories that cross a base cuboid in its X, Y, T dimension, respectively), which are used to tackle the distinct count problem.

```
declare
begin
    tdw.feed_tdw_mbr_bulkfeed('attiki');
    tdw.calculateauxiliary_cl('attiki');
end;
declare
begin
    tdw.feed_tdw_tbtree_bulkfeed('attiki', 'attiki_mpoints_tbtree_non_leaf',
             'attiki_mpoints_tbtree_leaf');
    tdw.calculateauxiliary_cl('attiki');
end;
```

Once a TDW has been built and filled in, OLAP queries are straightforward to be applied. Precisely, the user is provided with a number of operations (one for each measure of the fact table), in order to get aggregated information out of the TDW. The arguments for each one of the functions in the provided API (i.e. under the *aggregations* package) are: (in this order) the lowest and most left rectangle of the aggregated spatial area; the upper and most right rectangle of the aggregated spatial area, the first time period of the aggregated time period, and the final time period of the aggregated time period. For example, if a user wants to retrieve the total number of trajectories of an area starting from rectangle 370 and ending to rectangle 378 (see Fig. 12.24) and from the 4,645th to 4,650th h, she would have to execute the following query (Fig. 12.25):

```
select aggregations.aggregated_num_of_trajs(370, 378, 4645, 4650, 'attiki') from dual;
```

12.7 Progressive Explorative Analysis via Querying and Mining Operations

Typically in analytics, the user might decide to follow a progressive analysis approach. This section follows this path having as goal to explore interesting properties of our dataset via a representative set of mining methods. These mining

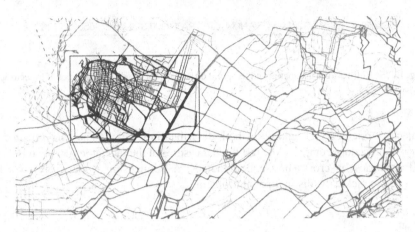

Fig. 12.26 A subset of the MOD as input to our explorative case study

operations include a number of algorithms for trajectory sampling, clustering, sequential pattern mining and others. Our progressive approach starts by selecting a number of trajectories from the initial dataset that corresponds to a region and a temporal period of interest. Subsequently, we apply a palette of techniques in a step-by-step methodology, while we often use the result from a step as the input to the next step. More specifically, our exploration starts by querying the initial dataset with a spatiotemporal range query that results in a set of 2,060 trajectories, shown in Fig. 12.26.

```
select * from table(
   tbfunctions.range(sdo_geometry(2003, 2100, null, sdo_elem_info_array(1,1003,3),
      sdo_ordinate_array(468828,4202769, 472367,4205112)),
      tau_tll.d_period_sec(tau_tll.d_timepoint_sec(2013,5,8,8,15,00),
                           tau_tll.d_timepoint_sec(2013,5,8,16,50,50)),
      2100,'attiki_mpoints_tbtree_non_leaf','attiki_mpoints_tbtree_leaf'));
```

Mobility data mining algorithms usually operate on a large MOD and it is natural for an analyst to wonder whether she could extract the same patterns operating on a much smaller and representative subset of the MOD. In other words, the question is rephrased to whether one can appropriately reduce a large MOD, taking only a sample of it, whose size is automatically computed, in an optimized and unsupervised way, and which encapsulates the mobility patterns hidden in the whole MOD. If the answer to the above question were positive, such a methodology would radically speed up several analysis and mining tasks. In the script below we follow such an approach by applying the T-Sampling algorithm (presented in Chap. 7), which samples the top-200 most representative trajectories from the 2,060 trajectories of the dataset produced by the previous step. The input and output datasets are illustrated in Fig. 12.27a, b, respectively.

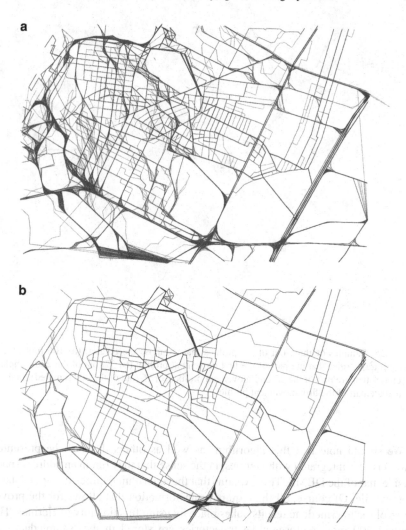

Fig. 12.27 Running the T-Sampling algorithm: (**a**) the input dataset (i.e. the result of the initial spatio-temporal range query at a different zoom-in granularity); (**b**) the top-200 most representative trajectories

```
declare
  value number;
begin
  value := hermes.odyssey_package.t_sampling(
    'tsampling_input.txt', --input dataset c:/hermes/io/ directory
    'sampledata.txt',     --output dataset
    5,     -- split temporal axis in 5 periods
    0.2,   -- distance tolerance for matching trajectories
    0.1,   -- epsilon, uncertainty threshold
    0.15,  -- sigma, Gaussian kernel parameter
    -100,  -- split space in 100x100 cells
    200);  -- top-200 representative trajectories
end;
```

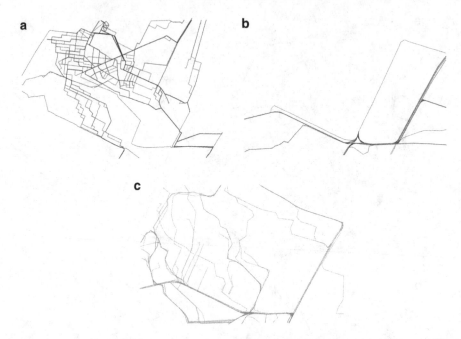

Fig. 12.28 Samples of three out of six profiles that compose the dataset: (**a**) 106 (out of 200 sampled) trajectories that are part of the first movement profile; (**b**) 37 (out of the 200 sampled) trajectories that are part of the second movement profile; (**c**) 56 (out of the 200 sampled) trajectories that are part of the fifth movement profile

We should note that this algorithm (as well as others that will be presented below) is soft-integrated with Hermes, in the sense that they run in an address space outside that of the DBMS. This explains that the algorithm requires its input dataset as a text file. Of course, all the required transformation that allows for the provenance of the intermediate results, takes place by using functionality of Hermes. The resulted 200 most representative trajectories are stored in the 'sampledata.txt', which subsequently it maybe reloaded to the MOD depending on the choice of the next analysis task.

Recalling the discussion in the beginning of the current chapter, these trajectories belong to different movement profiles. In Fig. 12.28, we illustrate three of the profiles, which almost compose the entire sampled dataset.

Next, let us try to find clusters of trajectories formulated in the sampled dataset. Initially, we apply the TR-FCM algorithm, which is the variant of the well-known FCM clustering algorithm, appropriately transformed for trajectories following a methodology as the one presented in Chap. 7 for making a partitioning clustering algorithm (as FCM is) applicable to trajectories. The following script invokes the TR-FCM via Hermes API.

```
declare
   value number;
begin
   value := hermes.odyssey_package.Tr_FCM(
      'tr-fcm_input.txt',      --input dataset c:/hermes/ directory
      'tr-fcm_clustering.txt', --clustering outcome
      0.05, -- Support threshold for considering regions to be dense
      5,    -- Split temporal axis to 5 periods (i.e. corresponds to the vector dimensionality)
      0.1,  -- Threshold to consider two regions (set of cells) similar or not
      0.0,  -- Uncertainty
      0.2,  -- A threshold that tunes the growing level
      -75,  -- Split space in 75x75 cells
      4);   -- K parameter (i.e. the required number of clusters)
end;
```

In our example, the 'tr-fcm_input.txt' contains the 200 most representative trajectories found by T-Sampling and 'tr-fcm_clustering.txt' contains the clustering results, which groups trajectories in four clusters, illustrated in Fig. 12.29a.

Next, we select one cluster found in the previous step by the TR-FCM algorithm (say, cluster 3) and we apply the Cen-Tr-I-FCM algorithm. Recall from Chap. 7 that Cen-Tr-I-FCM is an extension of the TR-FCM, which utilizes the Centroid Trajectory Algorithm (CenTra) instead of using an averaging method in its "update centroid" step.

```
declare
   value number;
begin
   value := hermes.odyssey_package.CenTr_I_FCM (
      'cen-tr-i-fcm_input.txt',      --input dataset
      'cen-tr-i-fcm_clustering.txt', --clustering outcome
      0.05, -- Support threshold for considering regions to be dense
      5,    -- Split temporal axis to 5 periods
      0.1,  -- Threshold to consider two regions (set of cells) similar
      0.0,  -- Uncertainty
      0.2,  -- Threshold that tunes the growing level
      -75,  -- Split space in 75x75 cells
      2);   -- Number of clusters
end;
```

File 'cen-tr-i-fcm_input.txt' now contains the 93 trajectories that belong to cluster 3 found by TR-FCM algorithm and we require to split it to two clusters, depicted to Fig. 12.29b (stored in 'cen-tr-i-fcm_clustering.txt').

Next, we follow a different strategy and we re-start the analysis of the dataset selected by the T-Sampling method by applying density-based techniques. The goal is to discover interesting properties of the dataset other than those found so far. We begin the exploration by utilizing the TRACLUS framework, which is a sub-trajectory clustering algorithm that ignores the temporal dimension of the

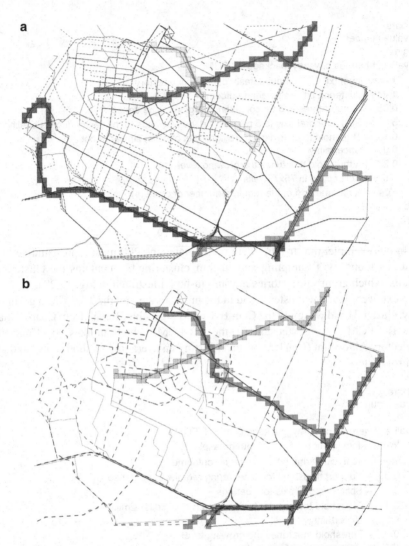

Fig. 12.29 TR-FCM algorithm in action: (**a**) trajectories grouped by TR-FCM in four clusters (depicted with different colors and patterns) and their centroid trajectories (differently colored cells); (**b**) splitting cluster 3 of the TR-FCM algorithm to two clusters discovered by the Centr-Tr-I-FCM algorithm

dataset. TRACLUS has been implemented in Hermes as a procedure, just like any other query. The following script applies TRACLUS to the top-200 trajectories stored at the 'traclus_input' table. The clustering outcome stored in the 'traclus_output' table is also illustrated in Fig. 12.30, where each cluster is visualized with a different color and line pattern as a thematic layer to OpenJump GIS (note that we do not depict the segments characterized as noise). The standard parameters of

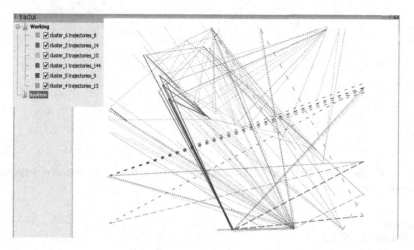

Fig. 12.30 Clusters found by the TRACLUS algorithm

TRACLUS, namely the *eps* distance threshold, the *MinLns* parameter for defining core segments and the smoothing parameter γ for computing the representative trajectory of each cluster are set to <1,000, 3, 0>, respectively. The next parameter of the algorithm is a flag, which is set to 2 if the default MDL-based approach is used to partition the trajectories to segments, while it is set to 1, if instead of MDL another compression technique is applied. In our case, we utilize the Douglas-Peucker line simplification method. If the latter is the choice, then the user should also provide the distance threshold parameter, which controls the level of simplification.

```
begin
    traclus.traclus('traclus_input', 'traclus_output', 1000, 3, 0, 2, null);
end;
```

In the following example, we choose two trajectories that move similarly (i.e. probably they would be grouped in the same cluster) and we demonstrate how we could generate a representative trajectory of them, by applying the TRACLUS's RTG algorithm. Recall from the discussion in Chap. 7 that a representative trajectory is an artificial trajectory that is extracted from a set of directed segments that move closely and towards (more or less) the same direction. Obviously, following this way we can discover the representative trajectory of each of the clusters produced by the TRACLUS algorithm. The script below, selects two trajectories (actually their sub-trajectories that correspond to the first Move episode), it produces a WKT file for their visualization and then it transforms them (with the use of the *segments_from_trajectories* method) from a collection of moving_point objects into a set of directed segments, namely a collection (i.e. a nested table) of unit_moving_point objects. Subsequently, the script applies the RTG algorithm and visualizes its outcome, which is illustrated in Fig. 12.31.

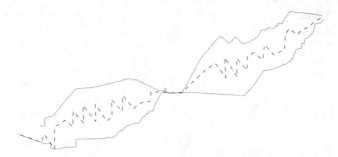

Fig. 12.31 Trajectories 1001 and 1185 (*solid lines*) along with their representative trajectory (*dashed line*)

```
declare
    srid integer:=2100;
    cur sys_refcursor;
    trajectories mp_array;
    segments unit_moving_point_nt;
    u_tab_t moving_point_tab;
    min_lns number := 2;
    smooth_factor number:= 200;
    representativetrajectory hermes.mp_array;
begin
    open cur for
        select p.sub_mpoint from attiki_sub_mpoints p
        where p.traj_id in (1001, 1158) and subtraj_id = 2;
    fetch cur bulk collect into trajectories;
    close cur;
    hermes.visualizer.movingpointtable2wkt(trajectories, 'mp_table.wkt', null);
    segments := traclus.segments_from_trajectories(trajectories);
    u_tab_t := traclus.RTG(segments, min_lns, smooth_factor);
    if u_tab_t is not null then
        representativetrajectory := mp_array(moving_point(u_tab_t, 100, srid));
        visualizer.movingpointtable2wkt(representativetrajectory, 'representative.wkt', null);
    else
        dbms_output.put_line('Could not formulate a representative');
    end if;
end;
```

Next, we apply the T-OPTICS algorithm, which has been enhanced with the choice of using alternative distance functions other than the original one that has been proposed. In Fig. 12.32a we illustrate the clustering result of the algorithm applied again to the top-200 sampled trajectories. The T-OPTICS is parameterized with *MinPts* = 10, *Eps* = 5,000, while we use as distance function, the Euclidean distance of the last points of the trajectories. Thus, this way we cluster trajectories w.r.t. their destination only. The final result is extracted by setting the reachability distance parameter to 451.15.

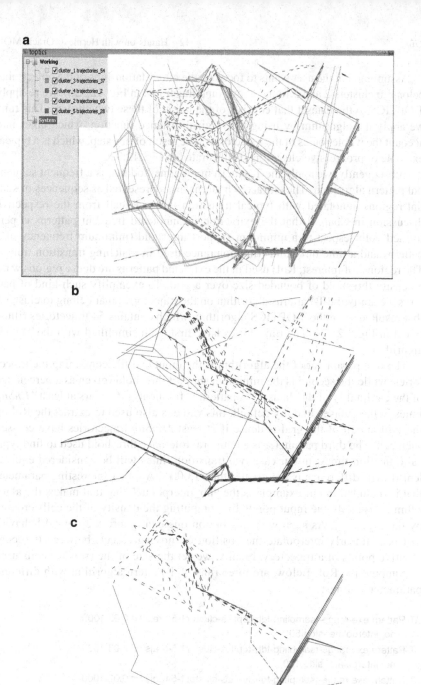

Fig. 12.32 T-OPTICS in action: (**a**) clusters found by T-OPTICS using the Common-Destination distance function (**b**) in particular, clusters 2 and 5 extracted by T-OPTICS using the Common-Destination distance function; (**c**) re-grouping clusters 2 and 5 with T-OPTICS using the similarity of the whole routes of the trajectories (noise is omitted from the illustration)

Assuming that the user wants to focus on the interrelationships of the objects that belong to clusters 2 and 5 found above (and illustrated in Fig. 12.32b), we re-apply T-OPTICS to the dataset that consists of the union of these two clusters. This time we apply the algorithm by using the original distance function, which takes into account the whole routes of the trajectories. The result of this step, which is a typical example of progressive clustering, is illustrated in Fig. 12.32c.

Subsequently, we apply the T-pattern mining method that is a frequent sequential pattern algorithm. The extracted patterns are represented as sequences of spatial regions annotated with typical transition times. Recall from the respective discussion in Chap. 7 that the extraction of annotated frequent patterns is performed with respect to a minimum support threshold (minimum frequency of a pattern) and a time threshold (time tolerance used in matching transition times). The regions (of interest, RoI) used in the extracted patterns are dense regions w.r.t. a density threshold of bounded size over a grid. To exemplify such kind of patterns we apply the T-Patterns algorithm on the trajectories that belong to cluster 1 that resulted from the T-OPTICS algorithm, which contains 54 trajectories (illustrated in Fig. 12.33a. The trajectories have first been simplified with the TD-TR algorithm.

The first parameter of the algorithm is the input text file containing the trajectories, while the second is the minimum support threshold (given as a percentage of the cardinality of the dataset). A pattern is frequent if it occurs at least *MinSup* times in the dataset, and, by default, this value is also used to extract the RoI on the grid: a cell of the grid is dense if at least *MinSup* trajectories have crossed such cell. The third parameter is a temporal tolerance threshold used to find typical transition times: in our case two transition times will be considered equivalent if their difference is not greater than 1,000 s. Another interesting parameter that is included in our example is the "no_interpolate" flag that makes the algorithm to use only the input points for computing the density of the cells crossed by the trajectory. As such with this option one can ignore the typical behavior that is to linearly interpolate the position of moving objects between two consecutive points of a trajectory. Finally, we set the size of the (square) cells used to compute the RoI. Below, are three runs of T-Pattern algorithm with different parameter settings:

```
.\T-Pattern.exe range-tsampling-tdtr-toptics-cluster1-54trajs.txt 0.03 1000
    -no_interpolate -side 50
.\T-Pattern.exe range-tsampling-tdtr-toptics-cluster1-54trajs.txt 0.03 1000
    -no_interpolate -side 150
.\T-Pattern.exe range-tsampling-tdtr-toptics-cluster1-54trajs.txt 0.03 1000
    -side 150
```

Figure 12.33b shows the discovered popular regions according to the first run, while Fig. 12.33c depicts again the RoI after invoking the algorithm with a larger cell size. Finally, in the last run of the algorithm we set off the no_interpolate flag and the result appears in Fig. 12.33d.

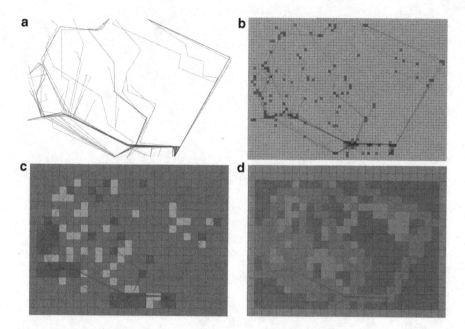

Fig. 12.33 T-Pattern in action: (**a**) cluster 1 found with T-OPTICS is given as input to T-Pattern; (**b**) T-Pattern found 34 ROIs and 512 frequent patterns; (**c**) cells are enlarged from 50 to 150, so 18 ROIs and 436 frequent patterns are discovered; (**d**) T-Pattern run without the no_interpolate flag be set and found only one RoI

Chapter 13
Hands-on with Hermes@Postgres MOD

Marine environment is very important in our society with implications in economy and natural preservation. Marine accidents that happen from time to time cost high amount of money to the involved parties and, more important, incalculable damage to the ecosystem. Unfortunately, there is lack of monitoring vessels at sea for conforming to the international marine safety regulations. To address this issue, the Automatic Identification System (AIS) had been made mandatory equipment for commercial and fishing vessels (above a threshold length). A vessel's AIS transceiver gathers data from onboard sensors like GPS and compass and broadcasts it publicly so that other vessels at sea or base stations at shore are aware of the position (latitude/longitude) as well as other trip-oriented information of the vessel under consideration. This chapter presents a showcase on a real AIS dataset and exploits the capabilities of Hermes MOD engine to efficiently process this kind of data.

13.1 Introduction: The Hermes@Postgres Data Type System

Hermes@Postgres[1] is a robust MOD engine designed to efficiently process moving object trajectories through a data type system that unifies the spatial and temporal dimensions of moving object trajectories. Table 13.1 presents the data types supported by Hermes, organized in three categories: temporal-only, spatial-only, and spatio-temporal data types.

It is clear from Table 13.1 that the data type system is rich enough to support spatio-temporal data objects that either have extent or not as well as their projections in time and space dimensions.

---

[1] The version of Hermes demonstrated in this chapter is the one built upon Postgres DBMS. This implementation is available for downloading for research and educational purposes under Hermes license at URL: http://infolab.cs.unipi.gr/hermes.

N. Pelekis and Y. Theodoridis, *Mobility Data Management and Exploration*,
DOI 10.1007/978-1-4939-0392-4_13, © Springer Science+Business Media New York 2014

Table 13.1 Data types supported by Hermes MOD engine

Category	Data types
Temporal	– Timestamp (datetime t)
	– PeriodT (timestamp initial, timestamp ending)
	– RangeT (interval radius, timestamp center)
Spatial	– PointSP (number x, number y)
	– BoxSP (PointSP lower-left, PointSP upper-right)
	– SegmentSP (PointSP initial, PointSP ending)
	– RangeSP (number radius, PointSP center)
Spatio-temporal	– PointST (timestamp t, PointSP p)
	– BoxST (PeriodT period, BoxSP box)
	– SegmentST (PeriodT period, SegmentSP segment)
	– RangeST (RangeT t-range, RangeSPsp-range)

Table 13.2 Top-10 flags of vessels included in the AIS dataset

Rank	Country	Number of ships
#1	Greece	263
#2	Panama (Republic of)	112
#3	Turkey	96
#4	Malta	76
#5	Liberia (Republic of)	32
#6	Saint Vincent and the Grenadines	29
#7	Russian Federation	28
#8	Antigua and Barbuda	24
#9	Italy	23
#10	Marshall Islands (Republic of the)	21

13.2   AIS Dataset Description

The dataset[2] covers almost 3 days of information about ships sailing at Greek Seas (mostly, Aegean Sea). In particular, time spans from "2008-12-31 19:29:30" to "2009-01-02 17:10:06" and space spans from (21E, 35N) to (29E, 39N) lon/lat coordinates. The number of ships included in the dataset is 933 using a variety of flags. Table 13.2 present the top-10 flags in our dataset, with the majority of them being "flags of convenience".

---

[2] The dataset used in this hands-on, called "IMIS3days", has been kindly provided to the authors by IMIS Hellas and is available for downloading for research and educational purposes at ChoroChronos.org.

Overall, the dataset is composed of 933 trajectories, consisting of 3,096,187 time-stamped locations in total, with the database table storing these records having a size of about 226 Mb. Figure 13.1 visualizes an overview of the dataset.

13.3   Loading the AIS Dataset into Hermes@Postgres

Loading a dataset into Hermes MOD engine turns out to be quite simple thanks to a module called "Hermes-Loader", which automates the creation and feeding of the respective database tables as well as the transformation of coordinates from degrees (lon/lat) to meters (x/y). The procedure is outlined in the following paragraphs, and from the presentation is clear the trajectory reconstruction (i.e. organizing data into trajectories) has to be done beforehand.

Before proceeding with the loader, one needs to save in PostgreSQL data directory the dataset in a file of type csv (with a header for column names) with the following structure: *<obj_id, traj_id, t, lon, lat>*. Then, the script listed in Fig. 13.2 runs in order to load the dataset into Hermes.

Query Q0. Loading the dataset.

1.   SELECT HLoader('imis', 'imis3days');
2.   SELECT HLoaderCSV_II('imis', 'imis3days.txt');
3.   SELECT HDatasetsOfflineStatistics('imis');
4.   CREATE INDEX ON imis_seg USING gist (seg) WITH (FILLFACTOR = 100);

In the script of Q0, lines 1 and 2 create the tables that will host the dataset and fill these tables, respectively, with the data that are found in file "imis3days.txt", according to the formatting discussed above. Line 3 calculates some statistics for the dataset, such as the average number of points per trajectory, the average duration of trajectories, and the average length of trajectories. Line 4 creates an index of type 3DR-tree on the dataset (recall the discussion in Chap. 4 about efficiently indexing moving object trajectories).

By default, the dataset is hosted in "imis_seg" table, according to a segment-oriented storage model (recall the discussion in Chap. 3 about alternative models for trajectories). The list of attributes of "imis_seg" is as follows: *<obj_id, traj_id, seg_id, seg>*, where *obj_id* corresponds to object's identifier (in our case, the MMSI of the ship), *traj_id* corresponds to a unique identifier of object's trajectory, *seg_id* corresponds to a unique identifier object's trajectory segment, and *seg* is the geometry of the trajectory segment, of type SegmentST (cf. Table 13.1). Figure 13.2 presents a few records of "imis_seg" table storing the dataset.

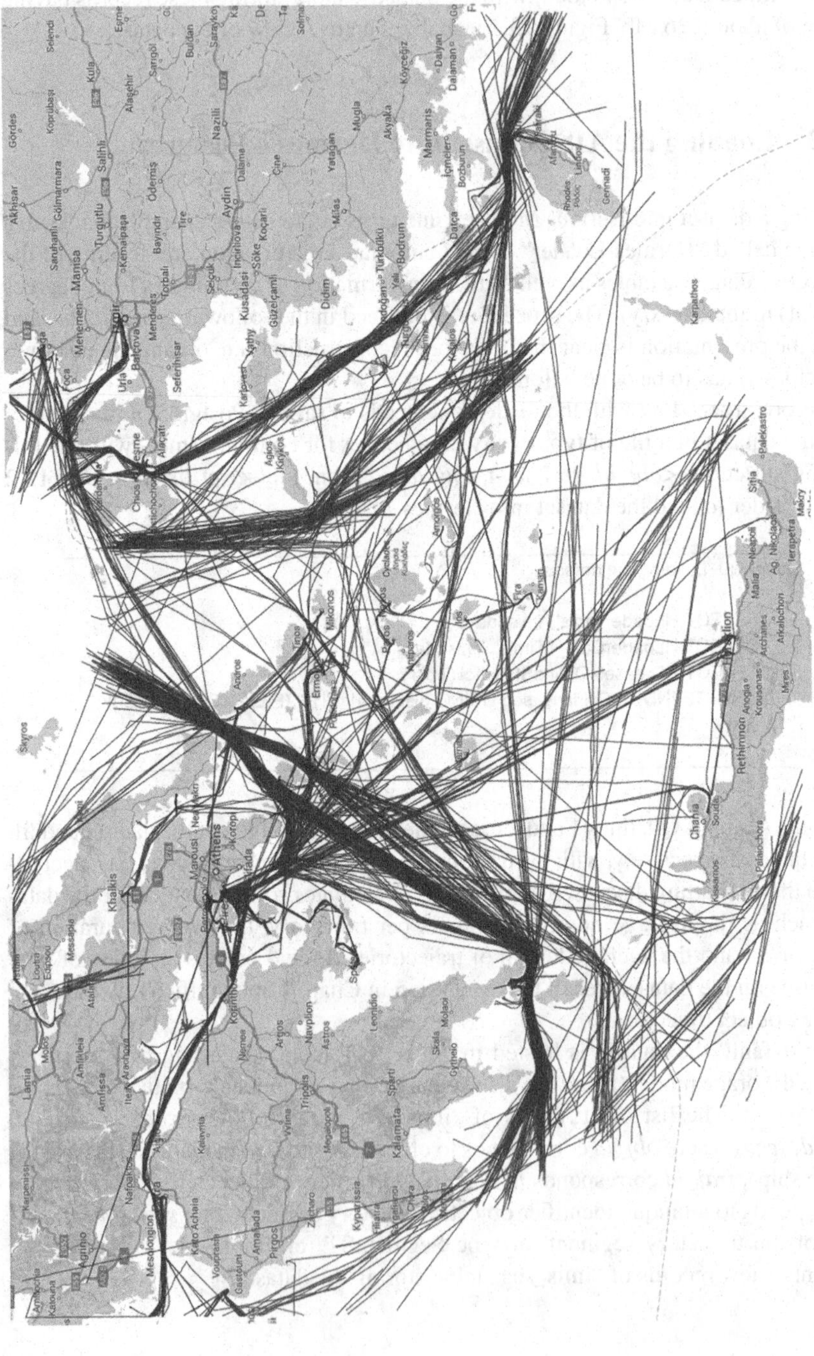

Fig. 13.1 Overview of the AIS dataset

	obj_id [PK] integer	traj_id [PK] integer	seg_id [PK] integer	seg segmentst
1	201100024	1	1	(('2009-01-02 08:54:07', (2739542, 4581334)), ('20
2	201100024	1	2	(('2009-01-02 08:54:25', (2739506, 4581230)), ('20
3	201100024	1	3	(('2009-01-02 08:55:06', (2739425, 4580996)), ('20
4	201100024	1	4	(('2009-01-02 08:55:56', (2739309, 4580726)), ('20
5	201100024	1	5	(('2009-01-02 08:56:16', (2739261, 4580623)), ('20
6	201100024	1	6	(('2009-01-02 08:56:47', (2739191, 4580459)), ('20

Fig. 13.2 A screenshot of "imis_seg" table

13.4 Querying the AIS Dataset

Once the dataset has been stored (and efficiently indexed) in Hermes, one can perform a number of queries exploiting on the strength of Hermes data type system (with a number of methods and operators on trajectory data). In the remainder of this chapter, we present several interesting queries on the AIS dataset using the API of Hermes. The images visualizing the results of the queries have been produced using "Hermes-Visualizer", a module of Hermes that allows constructing KML files built upon the results of the queries. A more technical description of "Hermes-Visualizer" API as well as example scripts in order to produce KML files are described in Sect. 13.5.

13.4.1 Timeslice, Range and Nearest-Neighbor Queries

Timeslice, range, and nearest-neighbor (NN) queries are essential for trajectory data management. In our case study, we present four queries of type timeslice (Q1), range (Q2, Q3), and NN (Q4). The difference between Q2 and Q3 is that the range in Q2 (Q3) is represented by a spatio-temporal rectangular (spatial circular, respectively) window.

– Q1: *"Find the positions of all ships at New Year's Eve 2009"*.
– Q2: *"Find the movement of ships inside the area of Piraeus port at New Year's Eve 2009"*.
– Q3: *"Find the ships that approached closer than half nautical mile to Cape Sounion, Attica"*.
– Q4: *"Find the ship that approached closest to Cape Sounion, Attica"*.

The SQL script that answers Q1 is listed below. In particular, the index-supported operator "contains" (symbol: ~) filters the database in order to select only those segments that contain the given timestamp "2009-01-01 00:00:00" in their temporal dimension. Then, the atInstant() method finds the exact location of objects at the given timestamp.

MMSI: 235060356

Fig. 13.3 Answering timeslice query Q1

Query Q1: timeslice query

```
1.   SELECT DISTINCT ON (obj_id, traj_id) obj_id, traj_id,
2.       atInstant(seg, '2009-01-01 00:00:00') AS position
3.   FROM imis_seg
4.   WHERE seg ~ '2009-01-01 00:00:00'::timestamp;
```

The visualization of the result appears in Fig. 13.3. As expected, we notice a lot of congestion in Piraeus port, which is the busiest port in Greece, with the main Aegean island ports (Heraklion, Rhodes, etc.) following.

Regarding range queries, using either a rectangular (Q2) or a circular spatio-temporal window (Q3), we exploit on the index-supported operators "intersects" (symbol:&&) and "within distance" (symbol:-<), respectively.

The SQL script that answers Q2 is listed below. Technically, the "intersects" operator (&&) filters the database in order to select only those segments that overlap the spatio-temporal window defined by interval "2009-01-01 00:00:00" ± 1 h in temporal dimension and two-dimensional rectangle with lower-left corner (23.59E, 37.91N) and upper-right corner (23.65E, 37.96N) in spatial dimension, bounding the area of port of Piraeus in lon/lat degrees (converted to x/y meters with PointSP() method). Then, the atBox() method finds the sub-trajectories within this range.

Query Q2: spatio-temporal range query

```
1.    WITH TO_METERS AS (
2.    SELECT
3.    PointSP(PointLL(23.59, 37.91), HDatasetID('imis')) AS low,
4.    PointSP(PointLL(23.65, 37.96), HDatasetID('imis')) AS high
5.    ),
6.    SPT_WINDOW AS (
7.    SELECT BoxST(
8.    Period('2008-12-31 23:00:00', '2009-01-01 01:00:00'),
9.    BoxSP((SELECT low FROM TO_METERS),
10.     (SELECT high FROM TO_METERS))
11.   ) AS box
12.   )
13.   SELECT obj_id, traj_id, (atBox(seg,
14.     (SELECT box FROM SPT_WINDOW))).s AS seg
15.   FROM imis_seg
16.   WHERE seg && (SELECT box FROM SPT_WINDOW)
17.   AND (atBox(seg, (SELECT box FROM SPT_WINDOW))).n = 2;
```

The visualization of the result appears in Fig. 13.4. There, we find that only one ship was docking at the port of Piraeus during that time period, which is reasonable since it was New Year's Eve.

Please note that atBox() may under certain circumstances return a (three-dimensional) point instead of a (three-dimensional) segment, such as when the intersection between the segment and the box is a point or when the segment and the period have only one timestamp in common. This explains why the method returns three properties: "n", "p", and "s". In particular, "n" informs whether the result is a point (value 1) or a segment (value 2) or there is no intersection between the segment and the box (value 0). Especially for values 1 and 2 of property "n","p" gets the point and "s" gets the segment, respectively.

On the other hand, the SQL script that answers Q3 is listed below. Technically, the "within distance" operator (−<) filters the database in order to select only those segments that overlap the spatial circular window defined by radius corresponding to 1.3 nm and center at (24.025E, 37.649N) corresponding to Cape Sounion, at the southernmost tip of the Attica peninsula. The nm2meters() function is used to transform nautical miles to meters, which is the adopted measurement unit for length in Hermes.

Query Q3: spatial-only range query

```
1.    SELECT DISTINCT obj_id, traj_id
2.    FROM imis_seg
3.    WHERE seg −< RangeSP(round(nm2meters(0.5))::integer,
4.        PointSP(PointLL(24.025, 37.649), HDatasetID('imis'))
5.    );
```

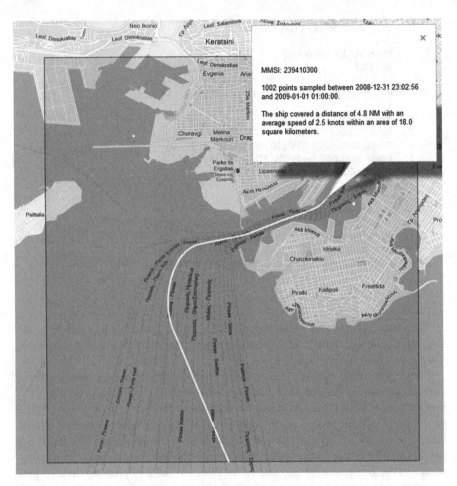

Fig. 13.4 Answering spatio-temporal range query Q2

    The visualization of the result appears in Fig. 13.5. There, we find a couple of
ships passing close to Cape Sounion.
    The SQL script that answers Q4 is listed below. Technically, index-supported
distance operator (<–>) selects the top-1 segment with respect to its distance from a
reference point (Cape Sounion: 24.025E, 37.649N). Notice that closestPoint() function finds the exact point within the trajectory segment that is closest to the reference point. Then, atPoint() function finds the timestamp corresponding to that point.
The value "false" on the last parameter of atPoint() enforces it to avoid checking for
containment since we already know that the point under examination is contained
on the segment. (In contrast, if the value of the third parameter is set to "yes" then
the cost of the calculation gets higher).

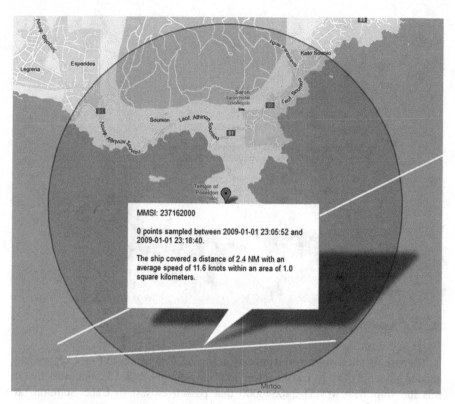

Fig. 13.5   Answering spatial range query Q3

Query Q4: NN query

1.   WITH TO_METERS AS (
2.   SELECT PointSP(PointLL(24.025, 37.649), HDatasetID('imis')) AS lighthouse
3.   )
4.   SELECT obj_id, traj_id, atPoint(seg, cp, false) cp,
5.   distance(cp, (SELECT lighthouse FROM TO_METERS)) AS dist
6.   FROM (
7.   SELECT obj_id, traj_id, seg,
8.   closestPoint(getSp(seg), (SELECT lighthouse FROM TO_METERS)) AS cp
9.   FROM imis_seg
10. ORDER BY seg <-> (SELECT lighthouse FROM TO_METERS)
11. LIMIT 1
12.) AS tmp;

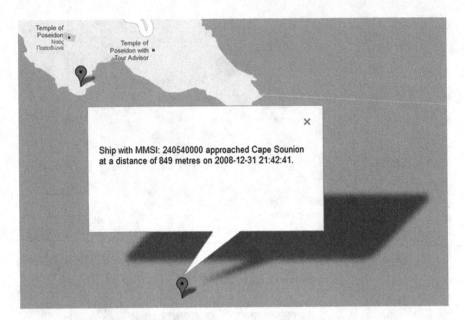

Fig. 13.6 Answering NN query Q4

The visualization of the result appears in Fig. 13.6. Whereas in Fig. 13.5, we had noticed a couple of ships passing close to Cape Sounion, Fig. 13.6 uncovers that it was the ship with MMSI 240540000 that passed closest to Cape Sounion (at 2008-12-31 21:42:41).

13.4.2   Join Queries

Join queries are also of great importance in trajectory databases since they join information from two datasets. In our case study, we exploit on self-join (vessel-to-vessel comparison) which sounds to be the most interesting:

- Q5: *"Find pairs of ships that were located closer than 1 nm at 2009-01-02 11:00:00 ± 5 min; for each pair, find their minimum distance of one ship from the other".*

The SQL script that answers Q5 is listed below. Technically, a timeslice query is executed first and its result is used by a (spatio-temporal circular) range query that follows; the former uses 2009-01-02 11:00:00 as the reference timestamp while the latter uses each ship's timestamp ± 5 min as temporal range and each ship's location ± 1 nm as spatial range. Moreover, the aggregate function HUnion() along with "GROUP BY" clause on the period component of the segments of a trajectory finds their union.

Output pane

| | Data Output | Explain | Messages | History |

	obj_id_1 integer	obj_id_2 integer	avg_dist numeric	common_period period
1	319287000	310076000	0.00672	('2009-01-02 10:55:00', '2009-01-02 11:05:00')
2	310076000	319287000	0.00779	('2009-01-02 10:55:00', '2009-01-02 11:05:00')
3	237088700	239642000	0.00973	('2009-01-02 10:55:00', '2009-01-02 11:05:00')
4	370058000	240846000	0.00998	('2009-01-02 10:55:00', '2009-01-02 11:05:00')
5	310076000	319127000	0.01031	('2009-01-02 10:55:00', '2009-01-02 11:05:00')
6	319127000	310076000	0.01048	('2009-01-02 10:55:00', '2009-01-02 11:05:00')
7	235059765	237671900	0.01248	('2009-01-02 10:55:00', '2009-01-02 11:05:00')
8	341579000	538002449	0.01303	('2009-01-02 10:55:00', '2009-01-02 11:05:00')
9	215821000	351426000	0.01466	('2009-01-02 10:55:00', '2009-01-02 11:05:00')
10	237671900	235059765	0.01489	('2009-01-02 10:55:00', '2009-01-02 11:05:00')
11	353778000	538002449	0.01493	('2009-01-02 10:55:00', '2009-01-02 11:05:00')
12	245184000	354407000	0.01657	('2009-01-02 10:55:00', '2009-01-02 11:05:00')
13	351426000	215821000	0.01667	('2009-01-02 10:55:00', '2009-01-02 11:05:00')
14	319287000	319127000	0.01676	('2009-01-02 10:55:00', '2009-01-02 11:05:00')
15	264900002	240712000	0.01707	('2009-01-02 10:55:00', '2009-01-02 11:05:00')
16	237096100	356813000	0.01716	('2009-01-02 10:55:00', '2009-01-02 11:05:00')
17	319127000	319287000	0.01718	('2009-01-02 10:55:00', '2009-01-02 11:05:00')
18	237015400	240344000	0.01727	('2009-01-02 10:55:00', '2009-01-02 11:05:00')
19	240712000	264900002	0.01728	('2009-01-02 10:55:00', '2009-01-02 11:05:00')
20	240344000	237015400	0.01740	('2009-01-02 10:55:00', '2009-01-02 11:05:00')

Fig. 13.7 Answering join query Q5

Query Q5: join query

```
1.  SELECT r.obj_id AS obj_id_1, db.obj_id AS obj_id_2,
2.      trunc(avg(metres2nm(distance(getSp(db.seg), getSpc(r.range))))::numeric, 5)
    AS avg_dist,
3.      intersection(HUnion(getT(db.seg)),
4.          Period('2009-01-02 10:55:00', '2009-01-02 11:05:00')) AS common_period
5.  FROM imis_seg AS db INNER JOIN (
6.      SELECT obj_id, RangeST('00:05:00', getT(position),
7.          round(nm2metres(1))::integer, getX(position), getY(position)
8.      ) AS range
9.      FROM (
10.         SELECT DISTINCT ON (obj_id) obj_id,
11.             atInstant(seg, '2009-01-02 11:00:00') AS position
12.         FROM imis_seg
13.         WHERE seg ~ '2009-01-02 11:00:00'::timestamp
14.     ) AS timeslice
15. ) AS r ON db.seg -< r.range
16. WHERE r.obj_id <> db.obj_id
17. GROUP BY r.obj_id, db.obj_id
18. ORDER BY avg_dist ASC;
```

A screenshot of the query result appears in Fig. 13.7. By examining it, we find that ships with MMSI 319287000 and 310076000 came too close to each other (in a distance less than 0.01 nm).

13.4.3   Topological Queries

Efficient processing of topological queries makes the difference for MOD engines with respect to spatial DBMS in the sense that the former are aware of the "trajectory" concept while the latter are only able to consider trajectories as polylines in three-dimensional space. In our case study, we present three different topological queries. Topological queries exploit on trajectory-based operations, such as *enter* (Q6), *cross* (Q7), and *start/end* (Q8).

– Q6: "*Find the ships that entered the port of Heraklion(irrespective of time)*".
– Q7: "*Find the ships that crossed Corinth Canal*".
– Q8: "*Find the ships that started their trip at the port of Piraeus and ended their trip at the port of Heraklion*".

The SQL script that answers Q6 is listed below. Technically, after defining a spatial box corresponding to the port of Heraklion, in the island of Crete, we exploit on the enter_leave() function. This function gets as input an array of trajectory segments and a box—in our example, the box corresponding to the port of Heraklion—and returns a structure consisting of two points: the enter and leave points of the trajectory with respect to the box (or value(s) NULL if there is no enter and/or leave point). In Q6, we are only interested in the enter point of the structure denoted by (el).enterPoint (see the WHERE clause at the final line of the SQL script).

Query Q6: topological (enter) query

```
1.   WITH TO_METERS AS (
2.   SELECT
3.   PointSP(PointLL(25.15446, 35.34621), HDatasetID('imis')) AS low,
4.   PointSP(PointLL(25.15664, 35.35251), HDatasetID('imis')) AS high
5.   ), PORT_AREA AS (
6.   SELECT BoxSP((SELECT low FROM TO_METERS),
7.   (SELECT high FROM TO_METERS)) AS box
8.   )
9.   SELECT obj_id, (el).enterPoint
10.  FROM (
11.  SELECT obj_id, enter_leave(array_agg(seg),
12.          (SELECT box FROM PORT_AREA)) AS el
13.  FROM imis_seg
14.  WHERE seg && (SELECT box FROM PORT_AREA)
15.  GROUP BY obj_id
16.  ) AS tmp
17.  WHERE (el).enterPoint IS NOT NULL;
```

The visualization of the result appears in Fig. 13.8. We find that the entrance points are in accordance with the predefined routes, which are drawn in the map background.

Continuing with topological queries, Q7 involves *cross* operator. In particular, we simulate *cross* operator by appropriately combining *enter* and *leave* operators. The SQL script that answers Q7 is listed below. Technically, after defining a spatial

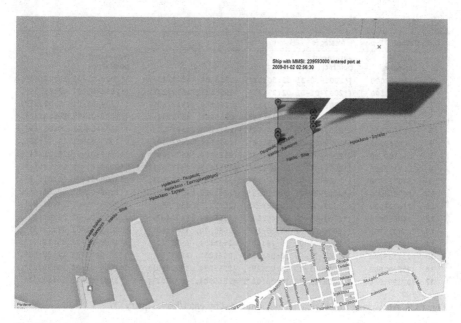

Fig. 13.8 Answering topological (enter) query Q6

box corresponding to Corinth Canal, the canal that separates the Peloponnesian peninsula from the Greek mainland, we again utilize the enter_leave() function; this time, we are interested in both enter and leave points, (el).enterPoint and (el).leave-Point, respectively, of the output of enter_leave(). In other words, we are looking for trajectories with a segment that enters the area and another segment that leaves the area under consideration.

Query Q7: topological (cross; simulated by enter / leave) query

```
1.   WITH TO_METERS AS (
2.   SELECT
3.      PointSP(PointLL(22.9599, 37.9173), HDatasetID('imis')) AS low,
4.      PointSP(PointLL(23.0117, 37.9532), HDatasetID('imis')) AS high
5.   ), CANAL_AREA AS (
6.   SELECT BoxSP((SELECT low FROM TO_METERS),
7.      (SELECT high FROM TO_METERS)) AS box
8.   )
9.   SELECT obj_id, (el).enterPoint, (el).leavePoint
10.  FROM (
11.     SELECT obj_id, enter_leave(array_agg(seg),
12.        (SELECT box FROM CANAL_AREA)) AS el
13.     FROM imis_seg
14.     WHERE seg && (SELECT box FROM CANAL_AREA)
15.     GROUP BY obj_id
16.  ) AS tmp
17.  WHERE (el).enterPoint IS NOT NULL AND (el).leavePoint IS NOT NULL;
```

Output pane

| Data Output | Explain | Messages | History |

	obj_id integer	enterpoint pointst	leavepoint pointst
1	239503000	('2008-12-31 22:30:11', (2555884, 4572743))	('2009-01-02 14:43:43',
2	237090100	('2009-01-01 02:27:37', (2555884, 4572620))	('2009-01-01 02:55:39',
3	240540000	('2009-01-01 04:25:11', (2561314, 4567750))	('2009-01-01 05:03:31',
4	237071200	('2009-01-01 04:35:16', (2555884, 4571598))	('2009-01-02 16:53:46',
5	341654000	('2009-01-01 06:20:58', (2561245, 4567750))	('2009-01-01 07:08:39',
6	341546000	('2009-01-01 09:35:16', (2555884, 4572636))	('2009-01-01 10:18:37',
7	232109000	('2009-01-01 15:27:46', (2555884, 4572579))	('2009-01-01 16:01:44',
8	215152000	('2009-01-01 18:07:56', (2561498, 4567750))	('2009-01-01 18:43:14',
9	240732000	('2009-01-01 19:08:38', (2561267, 4567750))	('2009-01-01 20:01:15',
10	667605000	('2009-01-01 20:15:51', (2561320, 4567750))	('2009-01-01 20:58:45',
11	240224000	('2009-01-01 21:38:37', (2561237, 4567750))	('2009-01-01 22:07:33',
12	240560000	('2009-01-01 22:40:54', (2555884, 4572467))	('2009-01-01 23:08:19',
13	273312900	('2009-01-02 05:55:30', (2561270, 4567750))	('2009-01-02 06:20:39',
14	240207000	('2009-01-02 09:25:40', (2555884, 4572731))	('2009-01-02 09:53:41',
15	237735200	('2009-01-02 10:04:52', (2561255, 4567750))	('2009-01-02 11:22:58',
16	355092000	('2009-01-02 13:07:29', (2555884, 4572669))	('2009-01-02 13:55:23',
17	372201000	('2009-01-02 13:08:01', (2555884, 4572732))	('2009-01-02 13:57:35',

Fig. 13.9 Answering topological (cross; simulated by enter/leave) query Q7

A screenshot of the query result appears in Fig. 13.9. According to this, a number of ships crossed Corinth Canal, a really busy canal.

On the other hand, the SQL script that answers Q8 is listed below. In this query, we follow a different approach in order not to repeat the usage of function enter_leave(), already demonstrated in Q6 and Q7. In particular, the process for addressing Q8 is as follows:

(i) spatial information about the locations of 100+ Greek ports is loaded (in table greek_ports);

(ii) the two ports of interest, Piraeus and Heraklion, are selected and, for each, a surrounding area of ±1 km distance is defined around the point locations, called s_area and e_area, respectively;

(iii) using these two areas and two appropriate operations, operator overlap (&&) and method contains(), the trajectory dataset is pruned to find those trajectories that, on the one hand, overlap the two areas and, on the other hand, have their starting and ending points contained in the respective areas, the port of Piraeus and the port of Heraklion, respectively.

Query Q8: topological (start / end) query

```
1.    CREATE TABLE greek_ports (
2.       name text NOT NULL,
3.       lon double precision NOT NULL,
4.       lat double precision NOT NULL,
5.       PRIMARY KEY (name)
6.    );


7.    COPY greek_ports(name, lon, lat) FROM 'greek_ports.txt' WITH CSV HEADER;


8.    WITH TO_METERS AS (
9.       SELECT name, PointSP(PointLL(lon, lat), HDatasetID('imis'))
10.          AS port_location
11.       FROM greek_ports
12.       WHERE name IN ('PIRAEUS', 'HERAKLION')
13.    ), PORTS AS (
14.       SELECT name, BoxSP(getX(port_location) - 1000,
15.          getY(port_location) - 1000, getX(port_location) + 1000,
16.          getY(port_location) + 1000) AS port_area
17.       FROM TO_METERS
18.    ), SE AS (
19.       SELECT S.name AS s_name, S.port_area AS s_area,
20.          E.name AS e_name, E.port_area AS e_area
21.       FROM (
22.          SELECT name, port_area FROM PORTS WHERE name = 'PIRAEUS'
23.       ) AS S, (
24.       SELECT name, port_area FROM PORTS WHERE name = 'HERAKLION'
25.       ) AS E
26.    ), START_END AS (
27.       SELECT obj_id, minT(i(seg)) AS start, maxT(e(seg)) AS end
28.       FROM imis_seg
29.       WHERE obj_id IN (
30.          SELECT DISTINCT obj_id
31.          FROM imis_seg, SE
32.          WHERE seg && s_area OR seg && e_area
33.       )
34.       GROUP BY obj_id
35.    )
36.    SELECT DISTINCT obj_id
37.    FROM SE INNER JOIN START_END ON
38.       contains(SE.s_area, getSp(START_END.start)) AND
39.       contains(SE.e_area, getSp(START_END.end));
```

A screenshot of the query result appears in Fig. 13.10. As indicated there, the dataset includes two ships (with MMSI's 23761100 and 239593000, respectively) that departed from the port of Piraeus and arrived at port of Heraklion.

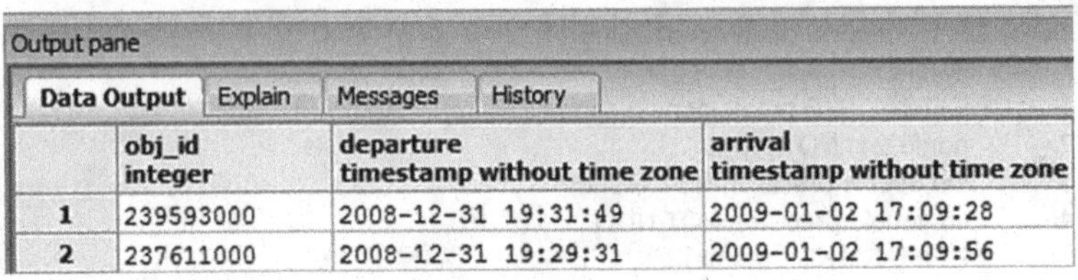

Output pane

Data Output | Explain | Messages | History

	obj_id integer	departure timestamp without time zone	arrival timestamp without time zone
1	239593000	2008-12-31 19:31:49	2009-01-02 17:09:28
2	237611000	2008-12-31 19:29:31	2009-01-02 17:09:56

Fig. 13.10 Answering topological (start/end) query Q8

13.4.4 Cross-Tab Queries

Cross-tab queries are, in general, more complex and more expensive than the previous ones. Nevertheless, they are very useful for analysis purposes since they provide deeper insight into the dataset under examination. In this section, we demonstrate two queries of this type.

– Q9: "*Perform equi-sized homogeneous partitioning in space and in time (e.g. a 10 × 10 grid in space and a 1-day interval in time) and count the number of ships per cell*".
– Q10: "*Taking ports of Greece into consideration, calculate the Origin-Destination (OD) matrix*".

It is clear that cross-tab queries like the above are expensive enough, since they involve a large number of calculations. In fact, in order to answer Q9, the trajectory datasets should be evaluated against a number of 300 (=10×10×3) spatio-temporal cells (recall that our datasets spans over 3 days in its temporal dimension). Even more expensive seems to the second cross-tab (Q10) since the number of Greek ports in our database exceeds 100, which results in an order of 100×100 matrix, hopefully sparse.

The SQL script that answers Q9 is listed below. In detail, we first create all 300 spatio-temporal cells, according to the following rule: triple (t_{id}, x_{id}, y_{id}), $1 \leq t_{id} \leq 3$, $1 \leq x_{id}$, $y_{id} \leq 10$, corresponds to one of the 3 days and one of the 100 spatial cells, e.g. triple (2, 5, 8) corresponds to values t in the second day, x in interval $(x_{min}+4\times(x_{max}-x_{min}), x_{min}+5\times(x_{max}-x_{min}))$, and y in interval $(y_{min}+7\times(y_{max}-y_{min}), y_{min}+8\times(y_{max}-y_{min}))$. Having set the partitioning, for each spatio-temporal cell, we execute a range query utilizing the overlap operator (&&) in order to find the ships that were located in the specific spatial area during the specific temporal interval.

Query Q9: cross-tab query

```
1.    SELECT t_id, x_id, y_id, count(DISTINCT obj_id)
2.    FROM (
3.       SELECT t_id, x_id, y_id, BoxST(ti, te, lx, ly, hx, hy) AS region_box
4.       FROM (
5.          SELECT row_number() OVER (ORDER BY ti) AS t_id,
6.                 ti, ti + '24:00:00'::interval AS te
7.          FROM (
8.             SELECT generate_series(tmin, tmax, '24:00:00') AS ti
9.             FROM (
10.               SELECT date_trunc('day', tmin) AS tmin,
11.                      date_trunc('day', tmax) AS tmax
12.               FROM HDatasets_Online_Statistics
13.               WHERE dataset = HDatasetID('imis')
14.            ) AS tmp
15.         ) AS tmp
16.      ) AS t CROSS JOIN (
17.         SELECT row_number() OVER (ORDER BY lx) AS x_id, lx,
18.                lx + length AS hx
19.         FROM (
20.            SELECT generate_series(lx, hx, (hx - lx) / 10) AS lx,
21.                   (hx - lx) / 10 AS length
22.            FROM HDatasets_Online_Statistics
23.            WHERE dataset = HDatasetID('imis')
24.         ) AS tmp
25.      ) AS x CROSS JOIN (
26.         SELECT row_number() OVER (ORDER BY ly) AS y_id, ly,
27.                ly + length AS hy
28.         FROM (
29.            SELECT generate_series(ly, hy, (hy - ly) / 10) AS ly,
30.                   (hy - ly) / 10 AS length
31.            FROM HDatasets_Online_Statistics
32.            WHERE dataset = HDatasetID('imis')
33.         ) AS tmp
34.      ) AS y
35.   ) AS regions LEFT JOIN imis_seg ON seg && region_box
36.   GROUP BY t_id, x_id, y_id;
```

Figure 13.11 visualizes the top-10 (out of 300) records according to the count of ships detected. It is clear that some areas are much busier than others.

On the other hand, the SQL script that answers Q10 is listed below. Technically, in this query we proceed as follows: (i) we calculate all combinations of the OD-Matrix (considering a 1 km distance around the center of each port); (ii) for each trajectory in the database, we find its starting and ending point; (iii) we join the two results according to the condition that the starting point of the trajectory is contained in the origin port and the ending point of the trajectory is contained in the destination port.

Fig. 13.11 Answering
cross-tab query Q9

Output pane				
Data Output	Explain			Messages
	t_id bigint	x_id bigint	y_id bigint	cnt bigint
1	3	4	8	285
2	2	4	8	234
3	1	4	8	182
4	3	4	7	121
5	2	5	8	110
6	2	4	7	108
7	2	3	4	107
8	2	5	7	105
9	2	4	6	101
10	2	4	5	95

Q10: cross-tab (OD-matrix) query

```
1.   WITH TO_METERS AS (
2.      SELECT name, PointSP(PointLL(lon, lat), HDatasetID('imis')) AS
3.         port_location
4.      FROM greek_ports
5.   ), PORTS AS (
6.      SELECT name, BoxSP(getX(port_location) - 1000,
7.         getY(port_location) - 1000, getX(port_location) + 1000,
8.         getY(port_location) + 1000) AS port_area
9.      FROM TO_METERS
10.  ), OD AS (
11.     SELECT origin.name AS o_name, origin.port_area AS o_area,
12.        destination.name AS d_name, destination.port_area AS d_area
13.     FROM PORTS AS origin INNER JOIN PORTS AS destination
14.        ON origin.name <> destination.name
15.  ), START_END AS (
16.     SELECT obj_id, minT(i(seg)) AS start, maxT(e(seg)) AS end
17.     FROM imis_seg
18.     GROUP BY obj_id
19.  )
20.  SELECT OD.o_name, OD.d_name,
21.     count(DISTINCT START_END.obj_id) AS nof_ships
22.  FROM OD LEFT JOIN START_END
23.     ON contains(OD.o_area, getSp(START_END.start))
24.     AND contains(OD.d_area, getSp(START_END.end))
25.  GROUP BY OD.o_name, OD.d_name
26.  HAVING count(DISTINCT START_END.obj_id) > 0
27.  ORDER BY OD.o_name ASC, OD.d_name ASC;
```

A screenshot of the query result appears in Fig. 13.12. In this cross-tab, we are able to identify the busy routes in the dataset. Actually, Piraeus—Heraklion—Piraeus turns out to be the busiest route in the database.

Fig. 13.12 Answering
cross-tab (OD-matrix)
query Q10

Output pane			
Data Output	Explain	Messages	History

	o_name text	d_name text	nof_ships bigint
1	PIRAEUS	HERAKLION	2
2	HERAKLION	PIRAEUS	2
3	PIRAEUS	SYROS	1
4	KORINTHOS	PATRAS	1
5	PATRAS	KORINTHOS	1
6	ELEFSIS	KORINTHOS	1
7	LAYRION	KEA	1
8	ELEFSIS	PATRAS	1
9	RAFINA	ANDROS	1
10	AEDIPSOS	ALEXANDROUPOLI	0

13.5 Visualization Tips

Hermes provides a set of functions that allow constructing a KML document within
a query in steps. In particular:

- function KMLPoint() returns a string that gives a KML point placemark element
  (with each point having the object and trajectory identifiers in its description);
- functionstring_agg() aggregates points;
- function KMLFolder() encloses aggregated points under one KML folder element.

By enclosing a folder element into a KML document element, a KML file is
ready to be written into a system file using "COPY" command.

To make this process as clear as possible, we choose the two first queries discussed in this Appendix, timeslice query (Q1) and range query (Q2) and enhance
their SQL scripts, in order to construct the KML files that were responsible for the
visualization of the results illustrated in Figs. 13.3 and 13.4, respectively. The
respective enhanced scripts for Q1 and Q2 are listed in Q11 and Q12, respectively.

Query Q11: visualizing query Q1 via a KML file.

```
1.   COPY (
2.   WITH TABULAR_RESULT AS (
3.   ------------------------------ Core Query -------------------------------------
4.       SELECT DISTINCT ON (obj_id, traj_id) obj_id, traj_id,
5.           atInstant(seg, '2009-01-01 00:00:00') AS position
6.       FROM imis_seg
7.       WHERE seg ~ '2009-01-01 00:00:00'::timestamp
8.   ------------------------------ End of Core Query ------------------------------
9.   )
10.  SELECT KMLDocument(KMLFolder('2009-01-01 00:00:00', string_agg(
11.      KMLPoint('MMSI: ' || obj_id, getSp(position), HDatasetID('imis')), ''))
12.  FROM TABULAR_RESULT
13.  ) TO 'C:\Program Files\PostgreSQL\9.2\data\Timeslice.kml';
```

Query Q12: visualizing query Q2 via a KML file.

```
1.    COPY (
2.    WITH TABULAR_RESULT AS (
3.    ----------------------------- Core Query ------------------------------------
4.      WITH TO_METERS AS (
5.        SELECT
6.            PointSP(PointLL(23.59, 37.91), HDatasetID('imis')) AS low,
7.            PointSP(PointLL(23.65, 37.96), HDatasetID('imis')) AS high
8.        ), SPT_WINDOW AS (
9.        SELECT BoxST(
10.           Period('2008-12-31 23:00:00', '2009-01-01 01:00:00'),
11.           BoxSP((SELECT low FROM TO_METERS),
12.               (SELECT high FROM TO_METERS))
13.        ) AS box
14.     )
15.   SELECT obj_id, traj_id, (atBox(seg, (SELECT box FROM SPT_WINDOW))).s
16.      AS seg
17.   FROM imis_seg
18.   WHERE seg && (SELECT box FROM SPT_WINDOW)
19.      AND (atBox(seg, (SELECT box FROM SPT_WINDOW))).n = 2
20.   ----------------------------- End of Core Query -----------------------------
21.   )
22.   SELECT KMLDocument(KMLFolder('Input area',
23.      KMLPolygon('Piraeus port area',
24.      BoxSP(PointSP(PointLL(23.59, 37.91), HDatasetID('imis')),
25.      PointSP(PointLL(23.65, 37.96), HDatasetID('imis'))),
26.      HDatasetID('imis'))) || string_agg(tracksFolder, ''))
27.   FROM (
28.      SELECT obj_id, KMLFolder('MMSI: ' || obj_id,
29.         string_agg(trackPlacemark, '')) AS tracksFolder
30.      FROM (
31.         SELECT obj_id, traj_id, KMLTrack(
32.            ------------------ Balloon Info ----------------------------------
33.            'MMSI: ' || obj_id || '<br/><br/>' ||
34.            count(*) - 1 || ' points sampled between ' || min(getTi(seg)) ||
35.            ' and ' || max(getTe(seg)) || '.<br/><br/>' ||
36.            'The ship covered a distance of ' ||
37.            trunc(metres2nm(sum(length(getSp(seg)))))::numeric, 1) ||
38.            ' NM with an average speed of ' ||
39.            trunc(mps2knots(sum(length(getSp(seg))) /
40.            extract(epoch from max(getTe(seg)) - min(getTi(seg))))::numeric
41.            , 1) || ' knots.'
42.            ------------------ End of Balloon Info -----------------------------
43.            , array_agg(seg ORDER BY getTi(seg) ASC), HDatasetID('imis')
44.         ) AS trackPlacemark
45.         FROM TABULAR_RESULT
46.         GROUP BY obj_id, traj_id
47.      ) AS tracks
48.      GROUP BY obj_id
49.   ) AS folders
50.   ) TO 'C:\Program Files\PostgreSQL\9.2\data\Range.kml';
```

Authors' Bios

Dr. Nikos Pelekis is a Lecturer at the Department of Statistics and Insurance Science, University of Piraeus, Greece. Born in 1975, he received his B.Sc. in Computer Science from the University of Crete (1998), his M.Sc. in Information Systems Engineering (1999), and his Ph.D. in Moving Object Databases (2002), both from UMIST, UK. His research interests include data mining, spatiotemporal databases, management of location-based services, machine learning and geographical informa- tion systems, whereas he teaches respective courses at under- and post-graduate level. He has been particularly working for almost 10 years in the field of Mobility Data Management and Mining, being the architect of the widely cited "Hermes" Moving Object Database (MOD) engine. He has offered several invited lectures in Greece and abroad (including Ph.D./M.Sc./summer courses at Rhodes, Milano, KAUST, Aalborg, Trento, Ghent, and JRC) on Mobility Data Management and Data Mining topics. He has co-authored more than 50 research papers and book chapters, while he is a reviewer in many international journals and conferences. He has served as Co-organizer of a EURO Stream on data mining and knowledge discovery (DMKD@ EURO'09). He has been member of the Organizing Committee for ECML/PKDD 2011. He is or was principal investigator for a number of EU-funded research projects (with FP7/MODAP, COST/MOVE, FP7/DATASIM, and FP7/SEEK being the most recent). For more information: <http://infolab.cs.unipi.gr/people/npelekis>.

Dr. Yannis Theodoridis is a Professor at the Department of Informatics, University of Piraeus, currently leading the Information Management Lab (http://infolab. cs.unipi.gr). Born in 1967, he received his Diploma (1990) and Ph.D. (1996) in Electrical and Computer Engineering, both from the National Technical University of Athens, Greece. His research interests include Data Science (management, anal- ysis, mining) for mobility data, whereas he teaches databases, data mining, and GIS at under- and post-graduate level. He is or was principal investigator for a number of EU-funded research projects (with FP7/MODAP, COST/MOVE, FP7/DATASIM, and FP7/SEEK being the most recent). He has served as general co-chair for SSTD'03, ECML/PKDD'11, and PCI'12, PC vice-chair for IEEE ICDM'08, member of the editorial board of the Int'l Journal on Data Warehousing and

Mining (since 2005), and member of the Symposium on Spatial and Temporal Databases—SSTD endowment (since 2010). He has delivered invited lectures in Greece and abroad (including Ph.D./M.Sc.-level seminars at Venice, Milano, KAUST, Aalborg, Trento, Ghent, Cyprus, and JRC Ispra) on the topic of Mobility Data Management and Exploration. He has co-authored three monographs and more than 100 refereed articles in scientific journals and conferences, receiving more than 1,200 citations (according to SCOPUS). For more information: <http://www.unipi.gr/faculty/ytheod>.

Printed in the United States
By Bookmasters